The Complete Guide to Foodservice in Cultural Institutions

The Complete Guide to Foodservice in Cultural Institutions

Keys to Success in Restaurants, Catering, and Special Events

Arthur M. Manask
with
Mitchell Schechter

JOHN WILEY & SONS, INC.

This book is printed on acid-free paper. ∞
Copyright © 2002 by John Wiley & Sons, Inc., New York. All rights reserved.
Published simultaneously in Canada.

Library of Congress Cataloging-in-Publication Data:
Manask, Arthur M.
    The complete guide to foodservice in cultural institutions : keys to success in restaurants, catering, and special events / Arthur M. Manask, Mitchell E. Schechter.
      p. cm.
    Includes index.
    ISBN 0-471-39688-5 (cloth : alk. paper)
    1. Food service management.   2. Restaurant management.   3. Caterers and catering—Management.   4. Food service management—case studies.   5. Restaurant management—Case studies.   6. Caterers and catering—Management—Case studies.   I. Schechter, Mitchell E.   II. Title.

TX911.3.M27 M326 2001
647.95′068—dc21

                                                                        2001026113

Printed in the United States of America.

10   9   8   7   6   5   4   3   2   1

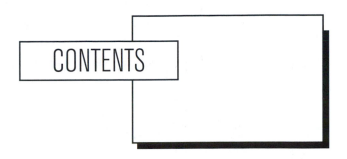

CONTENTS

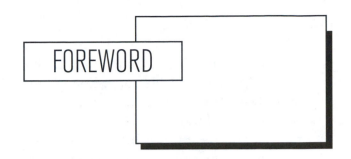

FOREWORD

Museums, by their very definition, are repositories of the world's greatest works of art and other treasures. This is no accident. Museums (and other cultural institutions) reflect the collective generosity, gifts, and talents of past and present-day individuals from around the globe, many of whom have dedicated their lives and careers to ensuring that the standards of excellence that have come to define modern museums are neither diminished nor compromised.

Museum-goers' sense of appreciation comes not only from engaging in the core activities of a museum visit—gazing upon an art object, participating in an educational presentation, attending a live concert that helps to interpret and enliven what's been seen—but also from other aspects of the visit, including experiencing the aesthetic and functional qualities of the building, participating in an on-site special event, and dining in an institution's restaurants or cafés.

Today, museums and other cultural institutions must not only strive to acquire and display premier collections, they must also earnestly endeavor to attain the same level of quality in all their foodservice operations and services. During a typical day at a museum, visitors may spend several hours viewing top-quality work by some of the world's foremost artists of all time. A pause for a meal or refreshments in one of an institution's foodservice facilities should, ideally, serve as a no less pleasant respite. Given this expectation, it is truly essential that by their quality, variety, and service, museum foodservices provide as much satisfaction, wonder, and joy as visitors have experienced in viewing the exhibits themselves.

Generally speaking, administrators at all types of cultural institutions face considerable competition for the resources necessary for daily operations, including collection management, exhibitions, acquisitions, educational programs, security, and facility administration. Nonetheless, it is increasingly clear today that the revenue that can be generated by a well-defined and successful foodservice and special-events program will contribute significantly to annual income available to underwrite and even expand cultural institutions' core activities. Correctly administered museum restaurants and special-events and catering services also generate goodwill and publicity, enhancing an institution's overall stature and importance in its community.

I believe that all of us who administer guest services in cultural institutions should study the contents of this book, as it can help us improve the quality and profitability of our current hospitality services, correct shortcomings, and successfully introduce new facilities and services. I commend author Art Manask for providing a long-needed reference work that can help administrators overseeing restaurants, special events, and catering at cultural institutions perform our jobs to our full potential.

Darrell R. Willson
Administrator
National Gallery of Art
Washington, D.C.

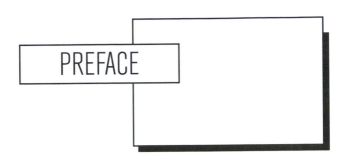

PREFACE

We set out to produce this book for several compelling reasons. The first is drawn from our understanding that business programs and missions have changed radically in recent years at most museums, as well as zoos, aquariums, and other public cultural institutions. This transformation has been driven both by rising visitor volumes and by guests' increasing expectations for service, which are challenging many administrators to maximize existing revenue-producing opportunities and seek to develop new ones.

We also believe that there are several other factors affecting the management of foodservice and facility rentals that administrators at museums and other cultural institutions need to see addressed. These include the fact that public funding has recently been reduced for many such institutions, requiring them to earn greater revenue from on-premise restaurants, catering programs, facility rentals, and special events. Increasingly forced to seek financial support from visitor retail purchases and from parties and receptions for corporations and community groups, a growing number of cultural institutions are looking for guidance to optimize their foodservice programs and create new or larger profit centers. Most administrators also now realize the importance of ensuring that all foodservice activities and facilities reflect their institutions' commitment to providing the finest possible visitor experience.

What this comes down to is that many administrators are now seeking assistance in their efforts to expand and enhance their overall visitor services. This is giving foodservice new significance as a means of "branding" and extending a cultural institution's reputation to new and returning guests.

Tied to these information needs is the fact that, to our knowledge, no other printed reference work exists to help museum and other cultural institution administrators successfully implement the program improvements they are now being compelled to make. We therefore decided to create a book that would help administrators manage the development, operations, renovation, and evaluation of foodservice programs at museums, zoos, aquariums, and similar cultural institutions. As you read through this book, you will also notice a selection of case studies that recount problems or shortcomings that foodservice programs have encountered and how these impediments were overcome to provide improved financial performance, management efficiencies, visitor satisfaction, and integration with institutions' missions and cultures. Our case studies present a cross-section of cultural institutions, including examples of "small" (less than 500,000 annual visitors), "medium" (500,000 to 1 million visitors), and "large" (over 1 million visitors).

All told, the goal we've set out to achieve is to help administrators at cultural institutions succeed in their professional responsibilities and respond to the growing need for progressively run foodservice operations. Our intention has also been to offer a work that you, as an administrator at a leading cultural institution, can refer to over and over again, to help resolve problems and evaluate the performance of your foodservice program. To this end, a glossary of commonly used foodservice terms and phrases has been included at the end of this volume. We hope you will find the experiences of your peers and our advice to be useful and effective in helping you oversee foodservices that both reach their full potential and express your institution's mission to your community.

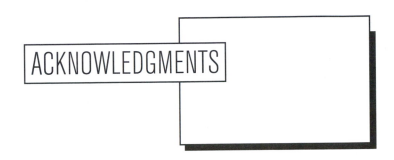

ACKNOWLEDGMENTS

I was introduced to foodservice and catering in cultural institutions in 1963, when I started work for my father and his foodservice company during the opening of the Los Angeles County Museum of Art's (LACMA) Plaza Café. The experience gained at LACMA proved invaluable when we opened the Garden Tea Room at the J. Paul Getty Museum in Malibu, California, in 1974.

Over a period of almost 20 years, we worked with museum administrators and professionals at LACMA and the Getty, including Stephen Rountree, Barbara Whitney, and Catherine Klose (who are still with the Getty Center), learning that operating museum cafés and providing catering services for special events in a museum setting pose unique challenges and opportunities.

After 25 years as an operator, I ventured into consulting. My first clients—Sherwood ("Woody") Spivey, at the Phoenix Art Museum; Darrell Willson, at the National Gallery of Art; and Tim Boruff, at the Indianapolis Museum of Art—encouraged me to expand our services to other cultural institutions in the months and years that followed.

Soon after completing the initial assignments that launched us in the unique world of offering consulting services to museums, zoos, aquariums, historic homes, and botanic gardens, I had the chance to work with Tom Mathews, at the Nelson-Atkins Museum of Art; Pat Grazzini, at the Minneapolis Institute of Arts; Warren Iliff, formerly of the Phoenix Zoo and currently chief executive officer at the Aquarium of the Pacific, in Long Beach, California; and Debbie Ives, at the Los Angeles Zoo. All these individuals and the ones previously named have been invaluable supporters and advocates as our practice and services have grown during the past eight years.

In 1999 Arthur M. Manask and Associates opened offices in Dallas and Chicago, and in 2001 we expanded to Massachusetts. I would like to thank our principals, Bill Heatley, Ray Sparrowe, and Eric Nusbaum, for their contributions to this publication.

I would further like to acknowledge Nancy Wright, at the Museum of Science and Industry; Teresa Sterns, formerly of the Science Museum of Minnesota; Laura Sadler, at the Field Museum; Sarah Christian, at the Denver Museum of Nature and Science; Susan Cole Bainbridge, Barbara Whitney Cole, and Suzanne Boué, at the Chicago Botanic Garden; Melody Kanschet and Stefanie Salata, at the Los Angeles County Museum of Art; John Easley, formerly at the Minneapolis Institute of Arts and currently at the Nelson-Atkins Museum of Art; and attorney Jeff Hurwit for their input and support in making this book become a reality.

I also offer a special thank-you to Susan Babcock Manask, JoAnna Turtletaub, of John Wiley & Sons, Inc., and coauthor Mitchell Schechter, whose support and efforts are sincerely appreciated.

I send thanks to the New York University student who sent me an e-mail several years ago asking if there were any reference works available on the management of restaurants and catering services in cultural institutions. It was this person, who said she had searched high and low for information on this topic on the Internet and university libraries and could find nothing, who originally motivated me to pursue publication of this reference work. And a very special thank-you goes to Ray Coen, who contributed to Chapter 6 and helped me develop the vision, mission, and marketing strategy for Manask and Associates, which has become the leading foodservice consultants to cultural institutions.

CHAPTER 1

Foodservice and the Visitor Experience

Why do millions of people in countries around the world flock every year to museums, zoos, aquariums, historic homes, and botanic gardens? While we can't state that a desire for memorable dining experiences or special events is an exclusive motivator, it would be wrong to undervalue the relationship between cultural education and hospitality services.

To us, it seems that even the basic activity of assembling and displaying selected man-made artifacts or items from the natural world is itself an act of hospitality, one extended by the organizing groups to the widest possible community and generations to come. Beginning with the collections, menageries, and fabled gardens created and displayed by rulers ever since antiquity, cultural displays have played a significant role in extending new knowledge about the world and humans' place in it. Improving the understanding of fellow citizens is an act of generosity and good intentions—hospitality in its very definition.

That the notion of offering more tangible nourishment (and an aesthetically pleasing and thematically coherent dining experience)

1

should naturally arise in environments established by cultural institutions is easy to grasp. What has proven more difficult for board members, directors, and senior staff, especially in our time of intense competition for consumers' disposable income and guests' increasing expectations for service and entertainment, has been the execution of foodservice strategies that successfully blend the objectives of an institution with its audiences' hospitality preferences and the needs of foodservice operators. Yet to operate institution-based restaurants, catering programs, facility rentals, and special events successfully, board members, executives, and administrators of service programs must regard foodservices as extensions of their institutions' core mission of informing and entertaining the public. Just as every cultural institution was founded and is operated to express a particular viewpoint, whether it be the meaning and history of the fine arts or the interrelatedness of elements of the natural world, so, too, should this purpose be applied to all aspects of visitor foodservice, from the simplest mobile snack cart to the grandest wait-served special event.

Museums and other organizations must also position their foodservice programs to meet their particular audiences' needs and expectations. We will address the logistics and processes by which this can be accomplished in subsequent chapters, but to begin to understand guests' foodservice likes and dislikes, programs should be viewed from three aspects—service (including management and planning), products, and facilities.

Most guests who visit a cultural institution's foodservices generally expect service that is polite, *quickly* responsive, knowledgeable, and offered in their own language. Younger and elderly guests, especially, may be unfamiliar with a cafeteria's scatter-system serving pattern or a café's self-service protocols, or they may have a hard time deciphering the choices presented by a multipart menu board. These kinds of visitor challenges, as well as the more sophisticated difficulties that can arise at catered and special events, need to be understood and addressed as proactively as possible by staff, managers, and senior personnel to ensure that guests receive an optimal service experience.

Due to the breadth and growing cultural diversity of the guest base served by institutions such as museums, zoos, botanic gardens, historic homes, and aquariums, foodservice menus need to be more diverse and creative than ever, though few items need to be rotated more often than seasonally to satisfy the needs of institutional staff, due to the relatively low frequency of repeat traffic. Today's guests want items that will be served quickly, can be eaten without formality (except at table-service facilities and events), appeal to at least two generations of family members, are fresh (not premade) whenever possible, and are both rich in flavor and relatively healthful. Food and beverage items should be of "high street" retail quality, and all merchandising and branding should be identical or equivalent to commercial standards, since neither your guests' taste buds nor their learned expectations degrade when they walk into your facility. In fact, since Americans now consume more meals away from home than in it, aspirations for food quality and choice may be even greater for cultural institution foodservices as, rightly or not, guests usually compare them to commercial counterparts.

Because many cultural institutions are housed in eye-catching spaces and/or grounds and excel at creating environments conducive to evoking particular visitor experiences, the development of appealing venues for visitor foodservices can and should be one of a program's strengths. It is useful for board members and senior staff to keep in mind, however, that many foodservice customers arrive at hospitality locations at least somewhat fatigued (and, one hopes, also exhilarated) after long walks perusing exhibits, so comfortable, ample seating and relaxing lighting and noise levels should all be high priorities when facilities are planned or renovated. Many visitors to cultural institutions also appreciate a chance to enjoy a break outside; therefore, patio, terrace, or courtyard seating is almost always popular. And, regardless of their check averages or level of decor, foodservice spaces offer excellent opportunities to remind guests of popular exhibits, reinforce an institution's purpose (through posters, theme menus, wall treatments, and the like), or allow access to a notable installation, such as a sculpture garden that can be viewed through a dining facility's windowed walls. In this way, guests are encouraged to remember why they have chosen to visit and are able to feel that they are still assimilating an institution's experience even as they are taking nourishment or refreshment.

Regarded in its entirety, the successful operation of restaurants, catering, and special events can do much to burnish—or tarnish—a cultural institution's reputation in its community. As the administrators cited in Chapter 3 point out, well-run hospitality programs and events can attract new clientele, daily revenue, donations, and bequests; forge productive relationships with new corporate sponsors; help to promote special exhibits and exhibitions; and generally increase visitor satisfaction with an institution's service offerings. However, these same programs and events can also, if presented indolently, indifferently, or even just inconsistently, diminish a cultural institution's renown and curtail the degree of support it finds in its community during fund-raising and membership drives.

The point is, the operation of your restaurants, catering programs, and special events should never be taken for granted; nor (as we will show) should you assume that your foodservices have reached their peak of either efficiency or creativity. Ensuring that each component of your hospitality program is judged first according to how appropriately it enhances guests' on-site experience will do much to keep these services on target. Our ageless interest in humankind's artistic expressions and the world's fauna and flora is complemented by our equally ancient desire to be welcomed, nourished, and made to feel as if our needs have been met with attentive concern. Cultural education and hospitality services, though not synonymous, overlap in their connection to our civilization's ritualistic roots, as they both offer us a means of creating order, even drawing inspiration, from the ineffably mysterious in life.

2

Understanding Foodservice Requirements in a Cultural Institution

The opening of a new cultural institution or the expansion of an existing facility provides an important opportunity to consider which types of foodservices would best serve an institution's visitors and coordinate with its mission. This chapter covers the factors that should go into an administrator's decision about what types of foodservice to offer in a new or expanded facility.

The first decision is whether a new foodservice operation is even a good idea. Administrators need to look carefully at the following questions:

- Why is adding another food/beverage location desirable?
- Have visitors (as well as staff and volunteers) been surveyed to determine if there really is a need for an additional facility?
- If the new facility is popular with visitors (including staff and volunteers) but is not profitable to operate, can it be eliminated without undue ill will?

■ Can a temporary form of foodservice be developed and situated where the permanent operation would be located, allowing a sort of trial run to see if the planned program meets identified visitor needs and financial operating criteria?

TYPES OF FOODSERVICE OPERATION

Once the need for a new operation has been established, the next decision is what type of facility it should be. Foodservices provided in cultural institutions such as museums, aquariums, botanic gardens, historic homes, and zoos typically are offered from one or more of a variety of facilities, including:

■ Restaurants

■ Cafeterias

■ Cafés

■ Snack bars

■ Coffee bars

■ Kiosks

■ Carts

Each type of operation has different characteristics and a different role to play in a cultural institution's foodservice program, as described below.

Restaurants

A restaurant consists of a separate (full or finishing) or front-of-the-house (display) kitchen and a fully enclosed guest seating area; it may also include a servery or ordering station(s), though table service is more common. A restaurant usually offers a full menu for one or more meals each day it is open, and most often its decor and furnishings are more upscale than any other foodservice operation in a cultural institution. Restaurants typically serve either members of the public or institution staff and their guests, and are frequently designed to host smaller special events.

Cafeterias

A cafeteria typically includes a full or finishing kitchen, a servery containing multiple points of service or food concepts (pizza station, grill, deli, salad bar, and the like), and a dedicated seating area. Serveries can be designed either with a single, continuous serving line or, more usually, in a scatter-system layout with separated, dispersed points of sale or food concepts. Some item pickup points will be self-service (salad bar, beverage station), while others will be staffed by servers and/or cooks (hot-entree station, grill). Cafeterias most often offer sufficient menu variety for customers to select one or more full meals during each day of operation. Decor and furnishings will usually be more functional and durable than aesthetically impressive, though theming these to cultural institutions' collections and exhibits will enhance the ambiance. Lately, cafeterias have begun to emphasize display food preparation at servery points of service, reducing space and equipment requirements for back-of-the-house kitchens. Cafeterias' dining and seating spaces are also used as special-event venues; at such times, servery areas are screened off while dish up and service take place. Cafeterias are usually open to both visitors and staff, though separate facilities of this type can be built to accommodate these customer groups individually. Cafeterias with separate, branded food concepts and/or distinct points of service are also referred to as "food courts."

Cafés

Simpler in concept and food selection than restaurants and cafeterias, cafés are essentially like a "Starbucks with a limited menu" of finger foods with coffee and noncoffee beverages, with seating provided.

Snack Bars

Snack bars, which commonly serve primarily visitors to a cultural institution rather than staff or members of the public, range in design from built-in-place "shops" to mobile units situated in busy trafficways. Rather than providing full meals, snack bars customarily offer items such as premade cold sandwiches and salads, baked goods, chips, fruit, candies, and ice creams, as well as hot and cold beverages. Typical equipment includes cold-holding units, merchandising racks or shelving, menu boards, and beverage dispensers. Customers serve themselves, and these snack bars almost always require no more than one or two staff to handle transactions and keep food and beverage displays and holding units fully stocked.

Depending upon their location, snack bars may be adjacent to fixed or temporary seating for up to a few dozen customers.

Coffee Bars

As the name suggests, coffee bars (which, like snack bars, range in design from built-in-place "stores" to mobile, temporary assemblies) are intended to offer a variety of coffee-based drinks, baked goods, and other light snack foods to members of the public and staff at cultural institutions. Product storage units, brewing and dispensing equipment, menu signage, and merchandising shelving or racks are all typically required in facilities of this type. Depending upon facility size and customer demand, coffee bars may require anywhere from one to several staff. This sort of bar is more likely than a snack-based facility to have themed decor reflecting a cultural institution's identity and ambiance, and to have dedicated seating adjacent to points of service.

Kiosks

Kiosks are usually temporarily or permanently assembled free-standing foodservice points of sale, typically located in high-traffic areas within or on the grounds of a cultural institution. Units of this sort are sometimes branded (e.g., Mrs. Field's Cookies). Because of their flexible design and "footprint sizes," kiosks can be used to display and sell a wide variety of items, including alcoholic beverages, grilled or heated foods, soups, baked goods, packaged snacks, soft drinks, coffees, and teas. A kiosk's labor requirements should not exceed two staff members. If hot foods or drinks are served, equipment can include roller grills, hot holding wells, microwave ovens, and brewing and dispensing units; otherwise, only storage units, cold holding cases, and merchandising racks or shelving will be required. Kiosks, unless of a "fixed" design, rarely have permanent utility or plumbing connections and usually do not have adjacent dedicated seating. Facilities of this type traditionally serve primarily museum visitors and not staff or members of the public.

Carts

Mobile by definition, carts are small, versatile points of sale that are usually situated wherever guest traffic is greatest. They are frequently used to augment built-in-place foodservice facilities during peak demand periods indoors and on outdoor patios. A cart's product assortment is limited only by the imagination of the foodservice provider and the tastes of institutional visitors and/or staff, but typically includes packaged snacks, baked goods, fruit, premade sandwiches, and salads and soft drinks in containers. More elaborate carts, with hot wells, grill tops, and equipment to brew and dispense hot beverages, are available but, when so equipped, these mobile service points will be larger, more difficult to relocate, and may require utility and plumbing connections. Due to their very low labor requirements, portability, and flexible product-display capabilities, carts can provide an acceptable, low-cost alternative to built-in-place or freestanding permanently assembled foodservice facilities. Depending upon their location, carts may be supported by adjacent customer seating.

QUESTIONS THAT GUIDE DEVELOPMENT

While different types of cultural institutions tend to have different mixes of foodservice options and thus different planning needs—for example, a museum or aquarium is more likely to consider a restaurant, café, or cart as part of a facility expansion, whereas zoos are more likely to include snack bars/stands and carts as part of their initial building project—there are certain key questions that need to be addressed by any institution during the planning process.

- If we build it, will the customers come?
- Is there a market for foodservice at this institution?
- Why is a foodservice facility being added?

- What are the capital costs associated with the project? Does the institution have to pay them, or can a foodservice operator be expected to front the build-out cost?

- What are the expected financial returns? Will the facility have to be subsidized?

- What are the tax implications (UBIT)?

- What impact will any new foodservice operation have on existing foodservice operations, catering, and special-events programs?

- Should the facility be operated in-house (self-operation) or outsourced to a contract management company or commercial catering firm?

PERFORMING MARKET RESEARCH TO ASSESS DEMAND

Performing market research is the best way for administrators to determine with any certainty whether a proposed new foodservice facility will attract customers. If an institution is adding a foodservice facility as part of an expansion or renovating an existing facility, the first step is to survey current visitors, members, volunteers, board members, and staff. If the facility will be located to afford the general public access without having to pay a separate museum admission, then local residents and/or members of the business community who might frequent the facility should be included in the research.

This market research can provide preliminary direction on probable users, concept, menu selections, hours of operation, style of service (such as table service versus self-service), size (number of seats and total square footage), and price points. Such research will also be helpful to the prospective facility operator (or manager, if the facility is to be operated in-house). In addition, it can help determine if an institution is a likely candidate for a branded restaurant concept; that is, a well-known local, regional, or national foodservice brand that in and of itself will draw visitors to the facility.

It is worth repeating here that market research should also be conducted on an ongoing basis as a way of measuring the effectiveness of the current foodservice program and of determining whether changes need to be made to ensure the highest level of customer satisfaction and the largest possible customer base and participation.

ASSESSING DIFFERENT TYPES OF FOODSERVICE

Among the foodservice operations that can be installed in nearly any cultural institution are:

- Self-service restaurant or café (may have a cafeteria- or food-court-style servery)
- Café or restaurant with table service
- Snack bar
- Coffee café (similar to a Starbucks)
- Kiosks or carts

Several different types of data go into the planning process. *Market research* can be expected to help determine which of these types of facility would best suit the expectations of an institution's visitors. The square footage to be devoted to foodservice is best determined by analyzing 12 months' worth of *visitor counts,* broken out by day of the week and by month. (If an expansion is planned, visitor count projections for the new addition should be taken into account, as well.) *Traffic patterns* will help determine where the facility should be sited; the best location is where visitors pass the facility at least twice, once upon entering and again when leaving. Such a location will ensure the highest possible participation level, since visitors have at least two chances to make a purchase. As well, a newly opened foodservice operation will be further enhanced if it is in a "free zone," or a location the public can access without having to pay admission. A free zone location will further increase visitor usage, as well as enhance the prospect that the facility will attract "restaurant only" customers from nearby residences and/or businesses. (If it cannot be placed in a free zone, customers can be given an identifying button or tag that permits free access to the dining facility.)

A new foodservice's visitor *capture rate* is likely to run between 20 and 35 percent, depending on where the facility is located. The 35 percent figure would be a safe estimate for a restaurant situated near an entrance, while 20 percent rate (or less) would apply to a restaurant that is not as convenient or easy to access. The capture rate can be used in conjunction with other data to help determine the size of the facility (see Chapter 8 for more information).

FINANCIAL CONSIDERATIONS AND JUSTIFICATION FOR A NEW FOODSERVICE

As a general rule of thumb, a cultural institution can expect its public restaurant or other foodservice facilities to generate from $1.00 to $3.00 per visitor in gross revenue. Why is there such variability? Determining factors include:

- The location of a facility within an institution. Is it located by an entrance, where all visitors walk by it at least twice, or is it in a remote location, where visitors will have to search for it?

- Is it in a "free zone," where entrance can be gained without payment of an admission fee?

- Can guests see and enter the facility without having to enter the institution itself?

- What is the demographic profile of the institution's visitors? A primarily adult group will generate higher per capita revenue than an institution that has children and school groups as a large percentage of its visitor base.

- If a well-known restaurant brand is being installed and fits well with visitor demographics, it is likely that the per capita revenue will be higher than if an unbranded restaurant was installed.

OUTSOURCE OR SELF-OPERATE?

The next step is to determine whether to outsource or self-operate the new facility. Most not-for-profit cultural institutions do *not* self-operate their foodservice programs, because running a restaurant or other dining facility tends not to fit well with an institution's primary mission. Therefore, it is advisable to prepare an estimate of income and expenses to determine whether the proposed foodservice will be financially viable. This can be done internally or by a knowledgeable outside consultant. If outsourcing is the desired direction, the request-for-proposal (RFP) process should be started as early in the planning stage as possible, with the goal of having the operator on board during at least part of the planning.

Most restaurants in cultural institutions are not big moneymakers, especially when operating costs and expenses such as utilities, building repairs and maintenance, heating and air-conditioning, and related services are factored in (such expenses are often not segregated or tracked). There is no precise rule linking annual visitor count to profitability—there are institutions with annual visitor counts of 100,000 to 200,000 or less that have profitable, self-sustaining restaurants, and there are institutions with 500,000 or more annual visitors that have unprofitable or marginally profitable foodservices. Generally, it is safe to assume that a well-located restaurant with high visibility and significant walk-by traffic will draw a larger visitor count, giving it a greater likelihood of financial success.

If a restaurant is not well located or is too small to handle visitor foodservice needs on its busiest days, then establishing secondary, temporary points of sale such as kiosks or carts should be considered. However, we do not recommend setting up permanent secondary points of service, as this risks cannibalizing sales from other, more potentially lucrative foodservice facilities. Also, while additional locations may increase overall foodservice revenue slightly, they can be very costly because of the attendant overhead (including labor, utilities, equipment repair and maintenance, janitorial services, and related out-of-pocket costs). Permanent secondary foodservice facilities should be considered *only* when a primary restaurant cannot adequately service the average daily visitor count.

REMODELING AN EXISTING RESTAURANT

In the restaurant industry, a rule of thumb is to reconceptualize a restaurant about every seven years. This same timeline should be applied to a restaurant located in a cultural institution. When a reconceptualization is being planned, it is useful to conduct market research among current restaurant customers and, equally important, visitors who are *not* currently patronizing the foodservice operation, to gain valuable input on which changes most need to be made in order to provide guests with a dining experience to which they look forward.

Dining trends and restaurant-goers' tastes continuously evolve, and administrators must be sure that their foodservice keeps up with these changes; otherwise customer counts and per capita spending will go down over a period of years. Administrators must never forget that visitors have a choice to dine in the institution or outside of it, and they must be sure to keep up with the many outside competitors looking to capture the same foodservice dollars.

When undertaking a foodservice remodeling, especially in older buildings, administrators should anticipate that it will take longer than planned and that the project may well go over budget, due to the many unknowns that will be encountered during and after demolition and during actual construction. Another consideration is the need to provide a temporary foodservice program during any major renovation. It is important to satisfy visitors' foodservice needs, if at all possible, during the period when a restaurant or other facility is closed. Temporary services can be offered from carts, fixed points of service set up in a gallery space, or even a tent (space and weather conditions permitting). If an institution's current foodservice routinely serves, say, 30 percent of daily visitors, the goal should be to serve the same 30 percent during remodeling, albeit with a more limited menu. It is usually sufficient to provide just hot and cold beverages, packaged sandwiches and salads, and bakery, dessert, and snack items. All of these foods can be prepared off-site and brought in daily. It is important, however, to have ample refrigeration on-site to ensure safe holding and service of perishable foods.

If a cart setup is selected to provide provisional service prior to a new permanent facility's opening, administrators need to interview several manufacturers that sell carts designed for public venues. Administrators should also interview users of these carts at institutions most similar to theirs to learn about likes and dislikes, what works and what does not, and what peers would do differently if they could do it again. Be aware that most local health departments have very strict rules and guidelines governing food-and-beverage cart operations to ensure the health and safety of customers. Administrators must be sure that they (or their operator) thoroughly research health department requirements, including storage and cleaning, before ordering or purchasing any mobile food-serving equipment. A cart manufacturer with extensive experience selling equipment appropriate for public attractions and cultural institutions can be a valuable source of real-world lessons and advice, but be certain to verify all information with the local health department.

CAPITAL AND START-UP COSTS, AND COMPOSITION AND FUNCTION OF THE DEVELOPMENT TEAM

The cost of building out a café or restaurant in a museum, aquarium, or other cultural institution can range between $150 and $300 per square foot ($+/-$), depending on the style, look, type of service, and quality of materials and finishes selected.

Early during the project planning process, the project team should include the building architect, a foodservice facility designer recommended by the architect or by the foodservice operator or operations consultant, an institution representative, and the outside foodservice operator (or manager if facilities are self-operated). Administrators need to be certain that the selected foodservice facility designer has had experience on similar projects.

If an institution's foodservice operator is a large regional or national restaurant/foodservice company, it is likely that this organization will provide the facility design, either generating it in-house or through a designer who is familiar with its restaurant concept(s).

Capital costs associated with the build-out of a museum's restaurant can be paid by the institution, by the operator (if outsourced), or shared. In most cases if the institution pays capital costs, the operator will be more interested in running the new facility. When an institution looks to an operator to put up all or part of a new facility's capital costs, this operator should be looked upon merely as a financing source—that is, when the facility is opened, all furniture, fixtures, and equipment (FF&E) will be owned by the institution, not the operator. When an operator provides the capital for a build-out, the total dollar amount is usually amortized over a ten-year period (assuming a low- to mid-six-figure investment as a minimum) with a buyback option. This means that in the event the operator's contract is terminated for any reason before the end of ten years, the institution will have to reimburse the unamortized capital investment (usually interest-free) unless it can pass off the obligation to a successor operator. Therefore, it should be an institution's goal, if at all possible, to provide the capital to build out a new foodservice operation as part of the overall building budget.

It is common for an external operator to provide the capital to purchase (and, in this case, own) the small wares and loose equipment, including point-of-sale (POS) system, required to run a new foodservice facility, as well as pay all preopening and start-up expenses and provide petty cash and change funds. If the facility is self-operated, these expenses will be incurred by the institution.

CONCEPT DEVELOPMENT: WHO'S RESPONSIBLE?

If management of a new foodservice is being outsourced, the operator will develop the overall program (including menus, style of service, presentation methods, merchandising, and other elements) subject to the prior review and approval of the institution. If administrators want a highly themed restaurant, then their institution can create the concept with the help of outside consultants who specialize in such development. Contract management and restaurant companies typically either have a collection of concepts among which administrators can choose or will be willing to develop a concept in conjunction with an institution, subject to prior review and approval by the institution. If the facility is being self-operated, the foodservice manager may, depending on his or her capabilities and the point at which he or she is brought on board, be able to establish the theme or concept. An outside consultant who specializes in concept development may also be hired.

EQUIPPING THE FRONT AND BACK OF THE HOUSE

Foodservice fixtures, furnishings, and equipment should be selected according to the proposed facility's concept, design, menu, and service program as developed by the project team (architect, foodservice facility designer, and operator or foodservice manager). But because, as noted above, all FF&E will usually belong to the cultural institution, and because it will usually remain in place for many years, it is important for administrators to maintain control and right of final approval over all such selections and budget.

The best choices are durable products with appropriate aesthetics that reflect the image of the institution. If a foodservice provider proposes any equipment or signage that is proprietary to the operator or its suppliers, administrators should be sure the contract with the outsourced company provides for the removal of these items when it leaves (at the institution's option). This will enable administrators or a successor operator to replace the equipment with items more suitable to a new menu program or food concept.

CONTROLLING ARCHITECTURAL INPUT

Architects are often most concerned with establishing the design statement a foodservice operation makes. However, it is important for administrators to be sure there is a balance between aesthetic needs and the practicalities of operating a foodservice facility of any type. This means that administrators should be at pains to listen to the restaurant operator, foodservice designer, and/or foodservice consultant during the design process.

ESTIMATING FINANCIAL RETURN

A rule of thumb for institutions with annual attendance of 500,000 to 750,000 is to expect a return from all foodservice operations ranging from 5 to 10 percent of total annual gross revenue. As annual attendance increases, higher gross percentage returns can be expected; for example, a museum with 1 million or more annual visitors can expect an average return from its foodservice operations of at least 10 to 15 percent—though it is important to note that the return on a restaurant or café will be lower than that on catering. (Different types of institutions can expect different rates of return. In zoos, for instance, the base return may be in the 10 to 15 percent range for annual attendance of 500,000 to 750,000 and approach 20 to 25 percent if annual attendance tops 1 million.)

If an institution provides the capital to fund a build-out of a new or expanded foodservice facilities, the annual percentage return will be higher; conversely, if the operator is required to provide the capital for build-out, an institution's percentage return will be lower. If a restaurant operator is also the institution's exclusive caterer, providing both food and alcoholic beverages, administrators are likely to see a higher overall percentage return. If the restaurant operator is only the exclusive alcoholic beverage provider (which is common) and the nonexclusive food caterer (whose presence is supplemented by a list of approved caterers), the return will not be as high (not counting income from approved caterers, if any) as under a totally exclusive arrangement.

Restaurants in cultural institutions with annual attendance below 500,000 tend not to be very profitable in themselves. However, when combined with catering opportunities, both internal (events for the institution) and external (served to outside groups and organizations that book events at the institution), a restaurant operation is able to spread fixed overhead such as management and chef's compensation over both programs, resulting in a more potentially profitable venture. It must be noted that there are a number of very small (under 100,000 annual visitors) cultural institutions in the United States that have very successful and profitable restaurants.

Profitability is a possibility when a cultural institution's foodservice facility becomes a destination restaurant on its own account.

The Ella Sharp Museum, in Michigan; the Nelson-Atkins Museum of Art, in Kansas City; the Denver Art Museum; the Bowers Museum, in Orange County, California; and the Arkansas Art Center in Little Rock are but a few examples of smaller institutions whose restaurants have become attractive destinations in and of themselves. Though many institutions would like their restaurants to become financially successful destinations that can support their institutions' missions and goals in regard to visitor services, it may be unrealistic to think that *any* cultural institution's restaurant can become a destination, though it is a good goal to strive for, as it helps ensure that the facility's operator or manager will work to achieve the highest food and service standards.

Though destination restaurants such as the ones mentioned above may be truly profitable, administrators should look carefully at the definition of profitability they are using. Oftentimes, an institution will determine profitability by subtracting from foodservice's gross revenue the direct costs of food, beverages, labor, and supplies without including utilities, bookkeeping and accounting, insurance, security, licenses and taxes, repairs and maintenance, janitorial supplies, telephone calls, pest control services, trash removal, armored-car transport of deposits, and other expenses, thus making a restaurant seem profitable when it is, in fact, subsidized by the institution. This is another example of why administrators must know how all costs are being accounted for.

When outsourcing, following is a list of the most common ways foodservice expenses are shared between operator and institution:

□ **Institution**	□ **Operator**
Utilities (gas, electricity, and water)	Food
Building repairs and maintenance	Labor and benefits
Furniture, fixtures, and equipment	Loose equipment
Rubbish removal (Dumpsters)	Operating supplies
Window washing	Light floor cleaning (dining, servery, kitchen)
Carpet shampooing; heavy floor cleaning in dining areas	Hood cleaning

Ceiling and light fixture repair
and maintenance

Pest control*

Flue cleaning

Local telephone calls*

Large equipment repairs,
maintenance, and replacement

Small equipment repairs
and maintenance*

Advertising and promotion*

There are three ways for administrators to determine what sort of commission or return from foodservice operations their institution should and can expect:

■ Engage an outside consultant to do a study

■ Talk to peers at similar institutions

■ Draft a request for proposal (RFP) to see what foodservice operators are willing to pay

While consultants who are knowledgeable about hospitality services in cultural institutions can offer advice and opinions, the best way to determine the financial potential of foodservice facilities at an individual institution is by initiating an RFP process that includes as many potential operators as possible and presents a variety of restaurant, catering, and service scenarios. Determining appropriate financial return through competitive bidding is the best and most accurate way to forecast projected foodservice income. If planning self-operation, an experienced foodservice manager and/or a qualified consultant can assist administrators with estimates and projections. Even if an institution's foodservice is self-operated, administrators may still be able to draft an RFP and explore the financial ramifications of outsourcing as an alternative. However, we must stress that we do not recommend an RFP process be conducted solely to check potential financial arrangements unless an institution is seriously considering outsourcing. Prospective operators are not likely to expend the time and money to submit a serious proposal (and it is not fair to ask them to do so) unless there is a real possibility that an institution will convert from self-operation to outsourcing if the financial and operational proposal so justifies.

* Items that can be provided by either party.

To some extent, administrators can benchmark or compare their foodservice facility's financial potential to those in similarly sized cultural institutions. However, while institutions may seem similar on the surface (having comparable annual attendance figures, for instance), when issues such as the location and type of foodservice(s), square footage of facilities, menus and menu pricing, visitor demographics, geographic location, outsourcing versus self-operation, number of points of sale, catering services, and alcoholic beverage service agreements are reviewed, administrators will find far more apples-to-oranges than apples-to-apples comparisons.

SELF-OPERATION

Self-operation of foodservices is found most often in zoos, less frequently in museums, botanic gardens, and aquariums, because self-operation historically has been part of zoo culture. Many cultural institutions that currently self-operate foodservices with annual revenues of around $2 million or less often times consider outsourcing alternatives at some point. Institutions that have higher gross foodservice revenue are more likely to be successful, both financially and operationally, at self-operation than are smaller programs—though, as noted above, there are exceptions.

The ultimate success driver for any foodservice operation, whether self-operated or outsourced, is the on-site foodservice management team, particularly the manager. In the case of a self-operated restaurant, it tends to be more difficult for the parent institution to attract and hold outstanding management talent, because, in a single institution there is no career path or growth opportunity equivalent to that which a manager would enjoy with a chain restaurant organization or contract foodservice company. Further, most cultural institutions do not offer the kind of profit sharing and other financial incentives that are commonly available in the private sector. For these reasons, it is common for a self-operated foodservice in smaller institutions to have a revolving door when it comes to the on-site manager's position. Larger institutions that can pay top dollar for their management (and culinary talent) tend to have fewer problems in this regard.

When Should a Cultural Institution's Foodservice Be Self-Operated?

Well-run self-operated foodservices can be more profitable (or less subsidized) than outsourced programs, because there is no third party earning a profit and thus all profits go to the institution. However, oftentimes an experienced outside operator can manage foodservices (especially smaller programs) more efficiently, purchase at lower prices, need fewer employees (because their proven operating systems are more labor efficient), and experience less staff turnover. The most market-savvy professionals are most often employed by foodservice contractors that can offer a cultural institution branded concepts and foodservice programs that may increase customer spending enough to cover the operator's profit, resulting in no loss of net income (and perhaps even a gain) for an institution, as well as a significant reduction in administrative hassles.

Where administrators choose self-operation, we strongly recommend using the services of professional foodservice consultants, who can provide guidance, input, audits, and assessments of foodservice performance on a regular basis. Most cultural institutions that self-operate their foodservices do not have senior administrators or accounting and auditing personnel who are familiar with the oversight of a foodservice department. While certain standard business principles apply, managing a foodservice business is very different from managing the primary operations and services of a cultural institution.

Mastering Manager Selection

Selecting and retaining the most qualified on-site manager is the most important and most difficult foodservice-related task for administrators at any cultural institution. It is important to ensure that an institution is offering a competitive financial package (salary, benefits, and bonuses—or regular salary increases if the institution's policies will not allow a bonus or incentive program). Administrators should look for managers whose experience in the foodservice industry best prepares them for the responsibilities at their particular institution. During the interview process, administrators should ask why the prospective foodservice manager wants to work at the institution, as opposed to pursuing a career in the commercial restaurant/hospitality industry. It is also important to find out how manager candidates perceive their future at the institution. Especially if there are no foodservice industry staff members at the institution with foodservice backgrounds, we recommend working with a search firm or using a foodservice industry consultant who is experienced at recruiting to assist with the interviewing and screening. After the candidates are narrowed down, be sure the first choice meets with and is interviewed by several department heads and others at the institution with whom he or she will be dealing on a day-to-day basis, to see how this candidate fits with the institution's culture.

Preparing to Add Staff

Foodservice managers at cultural institutions handle day-to-day subordinate staffing. Before an institution opens a new foodservice facility, its human resources department should become knowledgeable about issues relating to the recruiting, training, salaries and wages, and working hours and conditions of foodservice workers. These issues are often different for foodservice workers than for other employees at cultural institutions. Administrators, therefore, need to anticipate and budget for what is needed to support appropriate staffing.

OUTSOURCING—WHEN IS IT THE RIGHT OPTION?

Outsourcing allows a cultural institution to provide visitors, staff, and volunteers with necessary foodservices without having to undertake the management of these programs itself. While transitioning from self-operation to outsourced operation may cause some turnover in foodservice management and possibly some disruption, these difficulties can be outweighed by the long-term benefits. For example, an outside operator will handle all bookkeeping and accounting, human resources functions, purchasing, and related tasks. And while an institution might give up a little profit if it outsources its foodservices, in all likelihood it will get most, if not all, of it back through reduced employee turnover, better purchasing power, and improved training, systems, food merchandising, presentation, management procedures, advertising and promotional programs and lower administrative costs.

Prospective outside foodservice operators typically come from the following organizations:

- National and regional contract foodservice companies
- Local and regional restaurant operators
- Local hotels
- Local catering companies

Preparing for Operator Selection: The RFP Process

Cultural institutions can select an outside foodservice operator through an RFP process that may be either formal or informal.

The formal RFP approach includes the following steps:

1. Preparing a list of all prospective operators, grouping separately those that serve clients locally, regionally, and nationally
2. Advertising the business opportunity in national foodservice trade publications and through the state restaurant association
3. Sending letters to all prospective operators
4. Prequalifying prospective proposers
5. Preparing an RFP document that sets forth:
   a. An introduction and background
   b. The institution's goals and objectives for its foodservice operations (restaurant, catering, vending, alcoholic beverage sales, etc.)
   c. A description of the services to be provided
   d. Submittal information, including operator data (such as a financial statement and credit references), client references (current and former), proposed services (concept, menus, prices; if remodeling and renovation or build-out are desired, renderings, floor plans, layouts, and capital invest-

ment budget should be included), a financial proposal to the institution, a transition and/or implementation plan and timeline, proposed manager and on-site staffing and organization, contract terms and conditions, confidentiality terms, RFP schedule, and disclaimers

    e. Criteria for evaluating the submittals

6. Holding a pre-proposal meeting
7. Distributing questions and answers from the pre-proposal meeting to all potential providers
8. Conducting reference checks on finalist operators
9. Attending oral presentations and tastings by and question-and-answer sessions with finalists
10. Touring other, similar operations run by finalists
11. Interviewing finalists' clients (should be done by institution's administrators via personal phone calls)
12. Interviewing proposed on-site manager candidates
13. Selecting the preferred operator
14. Conducting negotiations with the preferred operator
15. Drafting a letter of intent
16. Handling contract negotiations
17. Preparing a contract document
18. Making a transition from the incumbent operator or self-operation to the new operator (who will start-up a new facility)

An informal RFP process, on the other hand, starts with a short list of prospective operators and then moves through most of the above steps on a one-on-one, negotiated basis.

Sourcing Operators Successfully

While it is difficult for a cultural institution to develop a comprehensive list of prospective foodservice operators, a good roster can be developed from the following sources:

- Peers at other institutions in the state, region, and nation
- Advertising in foodservice trade publications
- State restaurant association members

- Contacts in the local restaurant community found through board members and/or the local chamber of commerce

- Management consultants who specialize in developing foodservice RFP processes with similar cultural institutions

The more difficult task is determining which prospective operators have the experience, expertise, organization, and financial resources to become a good partner for an institution's foodservice. If administrators conduct a formal RFP process, requesting a token payment ($25 to $50) from all prospective operators who wish to receive the RFP is a good way to rule out those who are not really interested (or qualified).*

Usually only larger, more sophisticated foodservice operators will take the time and spend the dollars required to respond to a formal RFP process. Local restaurant and catering operators usually are not equipped to respond in this way but will respond if truly interested. If there is a local restaurant or catering operator that is a likely and worthy candidate to operate a cultural institution's foodservice facilities and program, engaging in direct negotiations that follow most of the above steps is an optional approach.

Common Contract Terms and Conditions

Contract options typically include either a management fee or a profit-and-loss arrangement. The following terms and conditions are applicable to both types of contract, though they may vary slightly when applied to either type, depending upon individual negotiations between the involved parties.

- The foodservice provider is to serve as an independent contractor.

- All foodservice staff are to be employed by the foodservice operator.

- All revenue is to be collected and all expenses paid by the operator.

- Conditions for termination are spelled out.

*Another approach is to have operators respond to an RFQ (request for qualifications) document. This will create an extra step, but if administrators are doing their own RFP process, this might help them decide which operators should and should not be on their list.

- Insurance requirements are listed.

- Both the operator's and the owner's (the institution's) operational responsibilities are delineated.

- Both the operator's and the owner's capital investment requirements are laid out.

- The owner retains the right to approve selection of the on-site foodservice manager.

- It is the operator's responsibility to secure all necessary business licenses and permits in its own name. (Depending upon state law and the wishes of involved parties, the alcoholic beverage license may be put in the owner's name.)

- The operator must remove any foodservice employee the owner does not find acceptable.

Key Differences in Contract Terms and Conditions

Profit-and-Loss Contract

- The operator pays the owner a percentage of gross revenue based on a sliding scale—a higher percentage at higher gross revenue levels; sometimes a minimum dollar amount is required from each program component (restaurant, catering, vending, etc.), with higher gross revenue creating higher commission percentages.

- The owner assumes no financial risk or administrative responsibilities for foodservice operations.

- The operator establishes all foodservice prices, with annual adjustments permitted subject to the owner's prior review and approval (with such approval not to be unreasonably withheld).

- The owner's auditing responsibility is limited to monitoring sales (revenue) to verify reported gross receipts and commissions received.

Management-Fee Contract

- The operator receives a guaranteed fee for its services, derived from a percentage of gross revenue (usually about 5 percent or minimum of ±$50,000 per year). The operator may also share in operational profits based on a previously agreed percentage split (usually 75 percent to the owner and 25 percent to the operator); sometimes, this split is made after the owner receives a predetermined percentage of net profits (typically 5 to 10 percent).

- The operator maintains detailed separate accounting books and records of income and expense.

- The operator pays all bills; invoices, payroll, taxes, insurance, and related charges are paid from this account.

- The institution assumes the primary financial risk.

- A portion (usually 10 percent) of the operator's fee is held "at risk" according to a mutually agreed-upon guaranteed annual budget provided by the operator and based on the owner's projected annual attendance.

- The operator recommends food and beverage selling prices, with annual adjustments permitted subject to the owner's prior review and approval.

- The institution's audit responsibilities include review and/or approval of complete profit-and-loss statements (all income and expense items). It is recommended that this audit be done on an annual basis.

CATERING AND SPECIAL EVENTS

External Catering

External catering is defined as those events staged at an institution by outside groups and organizations.

External catering is usually an institution's primary source of additional foodservice income. This income comes from facility rental fees, service charges (if applicable), and profits from food and alcoholic beverage catering (the difference between revenue and costs if the catering services are self-operated by an institution, or a percentage of gross revenue from food and alcoholic beverages if the catering is outsourced).

More and more institutions are actively marketing their facilities to outside groups and organizations. Many museums that are expanding are carefully looking at special-events needs as they relate to space planning, catering support, and equipment requirements, and are incorporating these criteria into expansion plans.

In order to optimize expanded special-events business from external sources, we recommend that individual institutions develop a marketing and business plan, have dedicated staff to market and sell the events, advertise in meeting planner and special-events publications, and pool advertising and promotion dollars with other local cultural institutions to showcase their community's cultural venues at special-events industry trade shows and via advertisements in industry publications.

Internal Catering

Internal catering is defined as day-to-day and special-events catering solely and exclusively paid for and sponsored by an institution. Whether catering is done by a single on-site or off-site caterer or by caterers on an approved list, an institution should negotiate to receive special pricing for internal events (10 to 20 percent off regular retail food-and-beverage pricing is customary) and a waiver of any service charges (other than gratuities that are 100 percent paid to foodservice staff) that are normally imposed. (Of course, the caterer would not be expected to pay a commission on such internal catering sales.) Such deals are preferable by far to "at cost" or "cost plus" arrangements with caterers, as these typically result in an administrative nightmare, not least because it is very difficult to determine what the costs really are.

Departments that order catering services will best understand their financial obligations if the caterer provides a price list with the discount already applied for day-to-day routine catering. Day-to-day catering for meetings, coffee breaks, and the like should be supported by a prepriced menu with the lowest pricing administrators can negotiate with the caterer(s). Because this type of catering offers the lowest profit margin for the operator and is offered as an accommodation to the institution, if an institution has several approved caterers, it is only fair to spread the event calendar among the entire list of caterers so that no one caterer suffers unduly.

When negotiating a discount for custom catering and special events such as luncheons, dinners, receptions, and the like, administrators should realize that, depending upon location and the type of event, caterers stand to make upward of 40 percent gross profit (that is, total food-and-beverage revenues less the cost of food, beverages, consumable supplies, and labor costs not billed to the user). Administrators' goal should be to receive a discount equal to about one-half of the caterer's gross profit. Under an exclusive catering relationship, this discount would be higher; the discount also tends to be higher in large cities than in smaller ones.

Establishing a Caterer Selection Process

In order to minimize administrative costs associated with managing and supervising caterers, we recommend that administrators maintain a short list of approved caterers that meet certain minimum standards. How many caterers would be on this list should vary depending on the city or community where an institution is located.

To begin the caterer selection process, the following should be undertaken:

- Identify all prospects for the approved list, including local caterers, hotels, and restaurants.

- Develop standards that each caterer must meet, such as having a kitchen approved by the health department and delivery vehicles capable of holding and delivering foods at appropriate temperatures.

- Note whether any of the caterers have experience with the institution or other local cultural institutions (this is always a plus).

- Determine if the institution wants a financial return from its caterers. It is customary for these caterers to provide a commission equal to about 10 percent of their total sales of food and alcoholic beverages (the range is from 8 to 12 percent, and percentages often differ for food sales and alcoholic beverage sales). Remember, the shorter the list of approved caterers, the higher the percentage an institution will likely be able to receive, since it is creating higher sales volume for a few caterers rather than spreading its sales volume over a long list. All caterers should pay the same percentages. This is important to ensure that one does not have a price advantage or disadvantage compared to others. If the on-site foodservice facility operator is one of the approved caterers, the percentage this firm pays should also be the same.

■ Ask for an annual minimum dollar commission if an institution is expanding a very active special-events program (i.e., there is a likelihood that each of the caterers on the approved list will be generating hundreds of thousands of dollars in annual sales). The purpose of a guaranteed minimum is to provide caterers with more of an incentive to bring clients to an institution, rather than just wait for telephone referrals from the institution on behalf of groups planning events. The minimum dollar guarantee, if one is adopted, can vary with each caterer and, in fact, can be part of the basis of selecting and narrowing the approved list.

■ Prepare an approved-caterer agreement. This should be a standard agreement, to be signed by all caterers. It should not be customized for each caterer.

■ Follow up by sending out an RFP to the list of caterers, setting forth minimum financial criteria and all other terms and conditions. This RFP should also ask for client references (particularly of value are those from similar institutions) and clearly state the evaluation criteria that will be used to get on the list.

■ Narrow the list upon receipt of proposals, then meet personally with the finalists to discuss the prospective relationship. Administrators should be sure the individuals with whom they meet are the ones they will be working with, not just sales representatives.

Weighing Exclusive Versus Nonexclusive Arrangements with Caterers

There are many factors that affect whether a cultural institution should have an exclusive relationship with one caterer or a nonexclusive relationship with a list of approved caterers. For example, if an institution has a restaurant that is likely to lose money or only break even, it is possible that the only way to secure an operator under favorable financial arrangements would be by agreeing to an exclusive catering relationship with that operator (assuming, of course, the company is also a qualified and reputable caterer). Further, if administrators are looking for a restaurant operator to provide a capital contribution to build, equip, and/or remodel, renovate, or expand a foodservice facility, they may have to give this operator exclusive rights to the likely more lucrative catering service.

In our experience, however, most institutions benefit from nonexclusive relationships. Nonexclusive relationships provide more upside potential for a greater number of special events if caterers can bring their clients to the institution's location and users can select the caterer that best fits their needs from among several on an approved list. The biggest negative of a nonexclusive relationship is that most cultural institutions are not designed or built for catering and special events, and even when they are, the use of five, eight, or more caterers substantially increases wear and tear even on the institution's facilities and creates additional administrative work for departments that interface with caterers, such as housekeeping, maintenance, and security.

Another factor to be evaluated when administrators consider whether to form exclusive or nonexclusive caterer relationships is whether board members, donors, or sponsors have particular relationships with caterers. When political issues such as this surface, which is very common at cultural institutions, it is possible to create two or three exceptions per year under an exclusive catering contract that will allow outside caterers to do business at the institution.

Exceptions Required for Kosher or Ethnic Catering

When an exclusive catering relationship is being contemplated, it is important that an institution specifies in the contract its right to engage the services of a kosher caterer or other ethnic or specialty caterer, regardless of whether this has been a past practice. This exemption can be covered in a contract in one of two ways: The institution or the outside group that requires a specialty caterer may contract directly with that foodservice firm, or the institution's on-site exclusive caterer agrees to handle the event by using a subcontractor approved by the customer based on an agreed upon pricing (markup) formula.

Dealing with Donated Food and Beverages

Most cultural institutions routinely receive donated beverages and foods. Therefore, catering contracts should anticipate and include this practice. If administrators have an exclusive caterer relationship, then this caterer should provide service and setup for donated food or beverages at cost or at cost plus an agreed-upon markup of about 10 percent. If administrators have a list of approved caterers, then each of these caterers should agree to provide this type of service on an equal basis during each year (i.e., each one agrees to stage up to two events each year with donated food and/or beverages).

Caterer Donations and Contributions

It is common for exclusive and approved caterers to provide a certain number of dollars annually based on the retail or at-cost value of catering services. Normally, the fewer caterers, the higher the dollar value received. Caterers may also, if asked in a request-for-proposal process or on a negotiated basis, offer a one-time or annual cash donation and/or purchase a corporate membership in addition to the catering services.

PRESERVING HEALTH AND SAFETY

Maintaining health and safety, though one of the most important aspects of foodservice operations, usually receives the smallest amount of visibility and attention.* Unlike displaying art or other collections or selling gifts and books in on-site shops, serving food and beverages to visitors, staff, volunteers, members, and guests is

---

*This information applies to both outsourced and self-operated foodservices. For additional information, see Chapter 11.

inherently risky if proper food handling, storage, and safety procedures are not in full force at all times. In most, if not all, cases, local city or county health departments will inspect institutions' foodservice operations. How often these inspections are conducted and how thorough they are is always a question to consider. In addition, foodservice managers will have their own inspection procedures. It is our recommendation, however, that cultural institutions retain the services of a qualified outside, independent inspection service. Many hotel and restaurant operators use this type of service to supplement their own internal inspection efforts and those of their local health departments as a double and triple check to safeguard their customers against food-borne illness caused by improper food handling and/or storage techniques.

Resolving Concerns About Food Safety Practices (A "Mini" Case Study)

Foodservice at a large museum was recording several million dollars in annual sales from several restaurants, catering, and special events. The programs were outsourced to a foodservice contractor, but the institution's astute administrators had concerns about the operator's food safety and sanitation practices, based on personal observations. Nonetheless, written reports from the local health department gave the operator at this institution a very high rating. How could administrators here resolve their concerns? Should they accept the health department reports or engage an outside service to perform an independent evaluation?

When it comes to the health and safety of an institution's visitors and staff, it is always best to err on the side of caution. Inspections from outside service firms that specialize in hotel and restaurant inspections are usually much more detailed and thorough than a local health department's. A local health department inspector might spend a couple of hours reviewing a foodservice, while an outside inspector might spend a full day or more going over the same facilities with a fine-tooth comb.

In this instance, administrators brought in an outside inspection firm that spent about three days on-site performing its initial inspection (the first inspection usually takes longer than subsequent reviews). Though the foodservice operation had passed the local health department's inspection a few days earlier, certain areas received ratings of 50 percent or less from the independent inspectors—which was very worrisome, considering that foodservice operators who support and maintain the highest food safety and sanitation standards routinely achieve minimum ratings of 85 percent on inspections by this firm, with most averaging in the low to mid-90s.

At this institution, the independent inspection spurred the foodservice operator to work with administrators to develop a plan ensuring that the highest possible sanitation standards would be maintained. The plan called for the outside inspection firm to conduct four reviews a year, which is the industry standard. The institution and the operator further agreed that the second inspection would result in scores of at least 60 percent (not including any serious health or safety issues requiring immediate attention) in all areas, while the third inspection's scores would be at least 75 percent and the fourth and subsequent inspections would result in scores of at least 85 percent. If any quarterly inspection fell below the agreed-upon scoring levels, the foodservice operator agreed to pay for that quarter's review. (For self-operated foodservices, the ratings should affect the manager's bonus, performance review or incentive compensation.)

The Role of Catering and Special Events in Membership and Sponsorship Development

Museums, zoos, aquariums, botanic gardens, and similar non-profit cultural education organizations all seek to inform and entertain different segments of the public, purposely targeting their attractions, resources, and promotions to specific interests and experiences in order to find the most receptive clientele. Yet all these institutions face common operational challenges, beginning with administrators' obligation to find new or enhanced sources of funding at a time when federal and many local governments are decreasing their financial support.

Naturally enough, when cultural institutions' boards, directors, senior staff, and other strategic planners set their minds on generating new revenue, increasing membership, creating greater use of on-site facilities, or boosting community involvement, maximizing hospitality services' potential is often a priority. But what are the most important specific expectations held by leaders of cultural institutions for catered and special events?

According to John Easley, director of development and external affairs for the Minneapolis Institute of Arts, "Our food, presentation, and services must match the quality and value offered by our entire museum resources, including our exhibits, if we are going to attain sufficient sponsor support and do all we can to enhance life in our community." In his senior administrative position, Easley has extensive involvement with his institution's hospitality programs. He believes that properly run catering and special-events programs support institutional growth and increase facility usage by two critical constituencies: current members, and current and potential corporate sponsors. "Our restaurants and special events provide important aspects of the museum experience for our 26,000 members," Easley stated. "We have one of the largest and best art collections in the country, and our members have made it clear that they don't want to experience any drop-off in quality when they take time to dine with us." When it comes to attracting corporate clients interested in staging catered and special events, Easley noted that most institutional administrators feel a heightened responsibility to provide impeccable services and foods, since corporate donors are counted on to underwrite special exhibits and installations. "It is vital that we do everything we can to accommodate corporate clients' individual requests or requirements, but we also try to encourage them to participate in achieving our goal of bringing art and art awareness to our community as a whole," he commented.

Catered events are held frequently at the Minneapolis Institute of Arts, including new exhibit preview parties for all levels of membership, thus giving administrators and service directors plentiful opportunities to raise needed new revenue from facility rentals. While Easley, among other museum administrators contacted for this book, acknowledged the importance of profits earned from these events, he added that they were put on to fulfill other objectives as well. "As it is throughout this institution, foodservice's first goal is to serve and please our current audience. But we also want to attract and build new audiences, and we've found that special events are highly effective ways to attract new guests to our rental spaces," he said. In addition, this administrator pointed out that when special events cover all their costs, as they do at the Minneapolis Institute of Arts, funds become available to support hospitality services departments and overall foodservice programs. Easley helps to set financial goals for foodservice and monitors its performance with his institution's chief operating officer, reporting to the museum's director and board only when a problem arises or a new facility or operator needs approval. "When it comes to finances, at a minimum we expect our foodservice program as a whole to break even," he stressed. "All of us here regard it as an essential amenity, one we must keep working to improve."

By forming a relationship with two private contract services companies when it outsourced its concessions, catering, and gift shop concessions programs in 1997, the Los Angeles Zoo was able to receive $1.5 million in new capital to improve foodservice and gift shop facilities. According to Debbie Ives, president of the non-profit group that supports the zoo (which is operated by the city of Los Angeles, its owner), revenue now being earned by foodservice's six fixed, themed facilities, as well as from catering and special events, provides critical ongoing funding for several areas of operation. For example, monies raised at donor events, such as the annual Wild Beast Society gala, help to support special programs, while a percentage of daily concessions and facility rentals are turned over to the zoo. "In addition, foodservice revenue helps to fund the zoo as a whole," Ives commented, "and facility rentals help us market ourselves throughout the community, which is extremely helpful, as our budget contains very little money for promotions."

This institutional executive has found that having day-to-day visitor hospitality services operate under the direction of a skilled and service-oriented general manager is a prerequisite for program success. Ives meets weekly with her concessions general manager, reviewing pricing, menus, and operational status; if Ives approves a requested program change, that decision is also referred to the zoo's director for approval. To ensure that all senior staff stay current on the operation of zoo restaurants, catering, and special events, changes in foodservices are always on the agenda when these administrators convene for their weekly meetings.

Ives further advised that the biggest challenge pertaining to the oversight of catered and special events at cultural institutions such as the zoo is "the correct allotment of management authority. At our zoo, until recently, the facility rental administrator did not work for the concessionaire, so coordinating party schedules, themes, and menus could be complex." In addition, at the Los

Angeles Zoo, as at many institutions, ongoing construction and maintenance programs often limit the number of on-site venues available for hospitality events. "What's more," Ives said, "all of our special events have to be conducted according to our charge of caring for and protecting the animals we have here."

Fostering respect for the living creatures on display and the environments that support them is also a primary priority at the Aquarium of the Pacific, in Long Beach, California. Here, the institutional mission is to instill in visitors wonder, respect, and a sense of stewardship toward the Pacific Ocean and its inhabitants, according to aquarium CEO and president Warren Iliff. One way this institution promotes its mission to guests is by theming its foodservice facilities. The main dining location at the Aquarium of the Pacific, for instance, is called Café Scuba and features displays of diving gear and environments. Table cards put out in the café also cite which fish are served here, which are not, and why, to emphasize the aquarium's commitment to ecological responsibility. "We have to coordinate and maximize all of our resources if we are going to be able to distinguish ourselves successfully as the aquarium of choice for divers and gear manufacturers in our visitor base," Iliff stated. Catering and special events are important revenue sources for this aquarium, according to Iliff, and include annual dinners provided for three founding donor organizations, which receive 20 percent off the rate this institution normally charges outside guests. "We also feel that our events help to expose the aquarium to more community members, especially older citizens, who are most likely to become personal or corporate donors," this executive pointed out.

Despite his senior role, Iliff finds it important to keep in close touch with foodservice operations, personally answering all written guest comments and having helped to select the current operator. He also works with the aquarium's chief financial officer to monitor daily operations and with the marketing department to promote concessionaire- and caterer-managed functions. As Iliff noted, "Special events can be key reputation builders, and those such as our annual fall and spring membership parties held in the Baja Pacific area of our facility really encourage guests to learn more from our exhibits, as well."

Administrators at many cultural institutions are finding that rising visitor volumes are compelling them to enhance or expand their food-based service programs. This situation is a familiar one for Suzanne Boué, director of visitor services at the Chicago Botanic Garden in Glencoe, Illinois, who saw guest counts rise some 35 percent in one year. Boué is responsible for foodservice, facility rentals, group tours, and all special events hosted for outside groups. To capitalize on the Garden's growing popularity, at the beginning of 2000 a rule requiring corporations seeking to hold special events to become corporate donors was rescinded, according to Boué. Currently, outside corporations may choose to become donors or not, and all still receive equal benefits on facility rentals. "Our rationale is that we are looking to maximize the number of special events we host, since they are a primary means of attracting and retaining the support of new guests and offsetting the recent decline in public funding," Boué related. "We're also looking to increase revenue generated by our visitor foodservice program by adding daily chef's specials in our main café and enhancing the overall quality of our foods and services."

Boué added that she firmly believes that positive experiences with on-site foodservices can encourage guests and corporations to rent the Garden's facilities for special events, but only if the operator's performance is commensurate with "the quality of our visitor experience here in the Garden as a whole." She also noted that despite special events' importance to her institution, they are never staged in such a manner as to compromise the Garden's aesthetics or detract from its mission of educating the public.

To ensure that these and other hospitality criteria are enforced, the Garden's senior staff and president are actively involved in foodservice administration. According to Boué, top executives here have a determining voice in regard to foodservice and special-event menus and pricing, and also closely track financial performance and guest feedback. "It takes a lot of time and attention to make sure your foodservice provider is responding as effectively as possible to growing demand and visitors' increasing expectations for a dining experience that's as high in quality as the rest of the Garden's attractions," Boué affirmed.

Achieving a successful balance between guest expectations,

institutional marketing objectives, and an operator's need for profit is very much the goal for guest service administrators at the National Gallery of Art, in Washington, D.C. "This museum receives only 50 percent of the funding it requires to put on exhibits from the government. We need to use hospitality services to attract the corporate sponsors we rely on for the rest of our revenue, since as *the* national gallery, we neither charge for facility rental nor admission," commented administrator Darrell R. Willson.

Many special events held at the National Gallery are elegant black-tie dinners attended by top-level politicians and jurists, designed to promote both the museum's fine arts collections and its reputation as one of the most prestigious hospitality venues in the capital.

Willson said that the most difficult challenge he faces as the executive officer of the gallery with oversight responsibilities for foodservice is "working with operators who come in here with a mind-set that is different from ours. What we at the museum want, more than anything else, are the highest levels of service, food quality, sanitation, and courtesy toward our guests. We recognize that operators must do no worse than break even by year's end, but what doesn't work for us is when they're chiefly concerned about profiting from every facility or event or are continually looking to improve their bottom line," he stressed. "We understand that it is difficult to operate restaurants that our visitors will rank among the finest of their kind in this entire city. However, if operators want to do business with cultural institutions such as ours, they have to keep in mind that maintaining our standards and consistently fulfilling all our guests' needs rank ahead of any other considerations."

When operated within appropriate parameters, Willson believes that foodservice can provide many benefits to cultural institutions. Private, on-site, wait-served dining rooms, for example, provide attractive locales where senior staff can discuss business with colleagues or entertain peers from other institutions, saving time and money that would otherwise be expended on outside dining. Such dining venues also give curators a place to host potential donors, while on-site catered and special events staged in conjunction with exhibit openings help to raise funds for future acquisitions, as well as defraying current operating costs for all museum departments.

Because the National Gallery's board must approve all corporate-sponsored special events, senior staff involvement in foodservices is extensive. Willson personally meets with his operator's foodservice manager and the gallery's director of special events to go over current operational performance, discuss any reported problems or complaints, and help plan upcoming events. This administrator also holds meetings each week with other gallery executive officers during which foodservices are regularly appraised. Each quarter, Willson participates in tastings of new menu items and meets with the president of his foodservice provider's organization to review financials and the current state of their relationship. Finally, foodservice renovation programs, such as the one completed during 2001 at all gallery restaurants, are discussed and approved by the institution's board.

At the Los Angeles County Museum of Art (LACMA), assistant vice president of special events and projects Stefanie Salata often finds that she is the museum's initial point of contact for members of the local community and corporations wishing to organize entertainments in her institution's facilities. Salata recognizes that her ability to accommodate special-event clients' needs, whether for affiliation of an event with a particular exhibit or a request for a one-of-a-kind menu, can help create relationships that lead to donations and philanthropic gifts to LACMA in years ahead.

"The revenue this museum gains from catered and special events helps us fund many activities, but the relationships they help us form and maintain are just as important," she affirmed. "The more guests and sponsors enjoy our special events, the more likely we are to reach our goal of being seen as a cultural asset to Los Angeles and its visitors."

Salata added that foodservice income is applied to the museum's marketing and operating costs, and also helps to reduce the institution's dependence on public tax money and fund-raising campaigns. While LACMA's cafeteria and restaurant already provide significant revenue, senior staff here continue to work with their foodservice provider to improve user-friendliness, pursuant to their ultimate goal of having guests see these dining locations as "destinations" that are as enticing as the museum's exhibits and installations. The same applies to special events, according to Salata. "We want guests to feel that being at a party or dinner at LACMA offers as much aesthetic pleasure as any of our artworks," she said. "Of course, putting on black-tie events is the most fun, but we've found that the long-term success of our special-events program is tied to our ability to serve each of our audiences at its own level."

With these administrators' goals, expectations, and caveats still fresh in mind, let's look next at some real-life foodservice problems, challenges, and opportunities encountered recently at cultural institutions, and how each of these were handled.

4

Achieving Self-Operation: Challenge-Solution Case Studies

Because the vast majority of cultural institutions operate autonomously and provide almost all visitor services through the efforts of skilled in-house personnel, it is logical that self-operation of dining and special events has been the norm until recent years. Yet as museums, zoos, aquariums, botanic gardens, and the like have sought to enhance their restaurant and catering programs in response to the rising expectations of guests, sponsors, and the general community, the traditional model of hospitality services has revealed frequently experienced shortcomings.

This chapter identifies a representative sampling of the sorts of difficulties actual self-operated museum restaurants and special-events programs have recently encountered and discusses how these problems were analyzed and which remedial steps were recommended to resolve them. It is likely that those readers who administer self-operated foodservices will recognize a past or present program shortfall among the following case studies; at the least, they may find themselves better prepared to make future program upgrades. We would also like to emphasize that while self-operated foodservices may encounter more or different problems than outsourced-managed departments, we in no way mean to denigrate or advise against self-operation of hospitality services at cultural institutions. In fact, in many cases, the ability of self-operated foodservices to meld with the culture and mission of a parent institution and to treat guests and staff alike as members of an extended institutional "family" may make this option preferable to outsourced management.

It has been our experience that, to paraphrase Leo Tolstoy, all problem-free foodservice programs are alike, but all troubled ones are troubled in different ways. Each cultural institution, of course, has its own expectations for its restaurants and catered events, and no such program can survive without winning the abiding approval of the organization's guests, donors, executives, and staff. Nonetheless, many self-operated hospitality services programs encounter significant difficulties, which may include the departure of key management or culinary personnel, changes in customer tastes, demographic shifts that require a redevelopment of menu or facility concepts, and shifts in a parent institution's mission that create a need for foodservice facilities to be "branded" or for special events to express the quality and aesthetic sophistication of an institution's exhibits or installations.

When a Senior Manager Departs

At a small art museum, the imminent retirement of a senior museum administrator who had personally overseen operation of the main visitor restaurant led to a comprehensive review of this facility's overall operations. Through on-site inspections, interviews with museum administrators and foodservice personnel, and survey-based assessments of customer satisfiers and dissatisfiers, it was determined that this museum's restaurant was in a precarious situation despite the fact that it was operating at a profit.

Besides the pressing need to reorganize the restaurant's management team, the facility's future profitability was in doubt due to the very low number of museum guests and local businesspeople who chose to patronize it (only 2.4 percent of the total number of this museum's visitors visited the restaurant, compared to an industry average goal of 20 to 30 percent). On-site research revealed that the large menu contained too many items that did not sell well, desserts were unattractive to contemporary tastes, portions were considered overwhelming by a significant majority of guests, and the style of service and tableware needed to be improved. On the financial side, it was found that food costs were above industry norms for a casual, wait-served restaurant; more specifically, a lack of formal procedures for menu planning, sales forecasting, recipe formulation, inventory auditing, product receiving, and purchasing were all exposing the operation to undue expenses.

ANALYZING THE RESTAURANT

As might be expected, the absence of written, formal operational practices was matched by an absence of hardware and software tools to facilitate smooth customer flows, handle credit card transactions, and plan, record, and forecast such key activities as ordering and storing supplies, producing menu items, and sales of all menu items. Further, on-site inspection had revealed that this restaurant's kitchen was overequipped and that several of its workstations needed to be reconfigured. Finally, analysis showed that in addition to increasing its sales, the restaurant needed to attract and host more special events and augment its support of the museum's community service mission.

Once these problems, concerns, and goals had been identified, the next step was for the outside consultant (which, as in all subsequent cases, was Manask and Associates) to present a report containing recommendations and options intended to address each identified current and potential problem area.

To increase traffic from this museum's total available customer base, we recommended that the restaurant seek to gain more patronage from proven market segments, in particular local community residents and museum visitors from outside the immediate area, by improving the menu.

The overabundance of menu choices and portions, which was causing unnecessarily inflated food and labor costs, as well as turning off some patrons, was addressed next. Recommendations included reducing the menu's daily soup choices from four to three and alternating selections, cutting baked goods choices from three to two and, again, alternating items, and deleting three of the least popular and/or least cost-effective sandwiches, two salads, and a fruit plate. Also suggested was a revision of the restaurant's dessert assortment to include lighter fruit sorbets (rather than ice creams), premises-baked goods, and dishes made with local fruit. Deleted menu items could occasionally be offered as specials. We advised that new plate presentation standards be established to emphasize foods' visual appeal rather than their quantity, that current tableware be replaced with more colorful contemporary dishes, and that beverages be provided from fountain dispensers rather than offered in cans.

To establish control of and effect eventual reductions in this self-operated museum restaurant's food costs, the following recommendations were prepared. First, the facility's management should begin tracking menu sales by item history, allowing more accurate production forecasting, and install food production management software to monitor inventory levels, match product purchasing to forecasts, resize recipes as needed, and calculate per-portion costs. Concurrently, a menu-engineering process should be initiated to ensure that the daily item mix (each day's combination of selections) would generate acceptable revenue levels while adhering to established cost parameters. The restaurant's managers were further advised that price increases might be necessary if high-cost items were to remain on the menu; the engineering process would determine if such adjustments were necessary.

To ensure that the restaurant functioned as efficiently as possible on a day-to-day basis, the introduction of a production planning cycle was advocated. This would ensure that sales forecasting would be done as needed, and that purchasing to the menu and standardized recipes and regulating production would be done in coordination with those forecasts. Supporting such management controls were recommendations to improve physical inventorying practices. Our advice included inspecting every delivered item before any were accepted into the facility, hand-examining products packed in cartons or cases, and weighing all items bought by weight on a scale to be purchased and installed near the receiving area, all to ensure that these items were as ordered. In addition, restaurant staff needed to start taking physical inventories of all stores on (at least) a monthly basis. This regular check would allow management to obtain and track the actual cost of goods sold in their restaurant.

To ensure that the restaurant purchased as cost-effectively as possible, its management was advised to form a "preferred vendor" relationship by contracting to buy all or nearly all foods and supplies from a single supplier, preferably a major wholesaler.

Operational improvements realizable from technological enhancements were addressed next. One change that we recommended strongly to this restaurant's management was the purchase and use of a silent paging system. Though usually only the largest or busiest restaurants require such a system, we endorsed its acquisition to reduce staff traffic into and out of a kitchen that was overequipped to the point of inefficiency. We suggested that the restaurant's customers, too, be given silent pagers, to allow restaurant staff to summon guests from museum exhibition spaces just when their tables became available. It also seemed likely that providing these pagers would both make time spent waiting more enjoyable for restaurant patrons, as well as help to drive more traffic through the museum's exhibits and installations.

Other potential technological upgrades offered for this restaurant's self-operators to consider included computerized order tracking and recording equipment coupled with credit card acceptance hardware. Such systems are available at various levels of complexity and cost, but the more functional and expensive examples tend to integrate more effectively with other automated management controls.

The next set of recommendations focused on the physical layout and overcrowded condition of the facility's kitchen. Suggestions began with having cooks tour their own work areas and remove any utensils, pots, pans, dishes, or other supplies and equipment not used regularly for meal production. These items, unnecessary shelf storage units, and even rarely used bulk dry goods could then be more appropriately (and securely) held in the restaurant's storage room. To alleviate this kitchen's particular layout problems, we advised that a stainless-steel worktable should replace the storage shelf units near the chef's table (and serve as a base for a small table mixer) and that a worktable near the range battery be replaced with a refrigerated sandwich makeup table with undercounter cold-storage units, allowing it to double as a salad prep station. Finally, it was pointed out that the elimination of several prep counters here would create room for a needed two-sink worktable, a baker's table, a standing mixer, a baker's proofing box, a convection oven, and a hand-washing sink for kitchen staff. Taken together, these changes were proposed to allow more effective production and presentation of the revised menu's choices, while potentially improving food safety, since more equipment and supplies would be located closer to where meals were actually being made up, thus reducing chances of product (cross-) contamination and loss of temperature control.

EVALUATING SPECIAL EVENTS AND FUTURE NEEDS

With systemic, operational, and physical plant concerns thus addressed, ideas were next provided to help the foodservice facility better support this art museum's mission and special-events program. We advised that a business plan be developed for the events program, to guide its overall development and to ensure that special events were in accord with this institution's stated intention of serving as the major cultural resource in its community. To be sure catered and special events were managed in a fiscally responsible way, we suggested that all foods for these occasions be purchased by the restaurant's kitchen manager; this would allow the museum to enjoy the savings engendered by the discounts available to its foodservice operation.

Finally, given the imminent departure of the senior administrator who had been acting as the restaurant's de facto general manager and the on-site management team's relative unfamiliarity with industry-standard practices and tools, it was necessary to evaluate the restaurant's leadership needs for the future. The best transition, in our estimation, was for the museum to hire an experienced and qualified full-time manager to oversee both restaurant operations and special events. Though the compensation package required to attract and retain a foodservice self-operator of that caliber might well absorb any existing surpluses from hospitality services, it seemed likely that only a seasoned individual with a comprehensive skill set would be capable of ensuring that existing surpluses would be restored and that the restaurant and special-events program would become more appealing and profitable in the years ahead.

ACHIEVING OPERATIONAL IMPROVEMENTS AND PROFITABILITY

When we turned our attention to the dining, catering, and hospitality services programs recently being self-operated at another, larger museum, we found ourselves considering a different (though related) series of concerns. After meeting with museum administrators, reviewing financial records and other documentation, and touring the museum's catering spaces, café, kitchen, storage areas, and staff dining room, the following foodservice goals and objectives were identified and agreed to:

- The museum's café and its catering service should achieve an annual surplus.

- The café should be a hospitality center for museum visitors and staff seeking daily snacks or lunches, as well as dinner on Wednesday and Friday nights.

- As a hospitality center, the café should appear uniformly clean, offer service to customers as quickly as possible, and serve fresh and healthful meals that were diverse enough in variety and price to appeal to everyone from students to businesspeople entertaining clients.

- The café should provide catering for events of up to 100 people, augmenting this museum's development and membership program, and support the largest (and largest number of) special events consistent with the capabilities of its kitchen, staff, and other resources.

Assembling an appropriate improvement plan for this self-operated foodservice and special-events program involved reviewing practices, conditions, and attitudes, and working toward change in those areas that seemed most problematic.

Though the museum's café had catered small and midsized special events for several years, hired a "chef," and developed and conducted activities with the museum's special-events coordinator, many museum staff still had reservations about the restaurant and its services. Specifically, despite the special-events department's willingness to use the café's catering services as often as possible, staff questioned whether the facility and its managers were capable of providing the necessary food quality and variety, and the service excellence required to present events as appealing as those staged by an outside catering specialist.

We also discovered that the museum needed to reconceptualize the management of hospitality services. At the time of our evaluation, an outside caterer was being hired regularly to handle larger special events, alcoholic beverages were being purchased by the special-events department and set up and served by outside catering staff, and the vending machine program was being administered by the institution's purchasing department. Also, the special-events department would establish temporary foodservice sales points to capture traffic drawn by new or highly popular exhibitions, usually employing the assistance of an outside caterer, under the reasoning that the café's management and staff were at pains just to handle the extra-heavy restaurant volume generated by these special exhibits and thus could not be counted on to provide sufficiently high-quality products and services at additional remote locations.

In addition to the multiple management centers and means of providing foodservices, hospitality-related record keeping was (not surprisingly) less than ideally organized. We found, for instance that the café had actually earned less revenue from visitors (who, paying full menu prices, recorded average checks of about $5.00) and more from museum staff (whose average checks were about $2.50, thanks to their in-house dining discount) than had been recorded. Perhaps more critically, customer traffic had not been analyzed in a way that would help this institution and its self-operated foodservice manage programs better or optimize operating efficiencies. This meant that customer counts by day-part or hour and menu "extracts" (the amount of each menu item served each day) were not being summarized and tracked, preventing the café's self-operators from accurately aligning food and labor costs with demand. Compounding this shortfall was the fact that production records, containing daily data on which menu items were produced and the amount of consumption and leftovers per item, were no longer being maintained by the café's management. In addition, portion control (the weighing and sizing of servings) was not being conducted according to formal, consistent procedures, and there was no ongoing staff training in such key areas as food safety, Hazard Analysis and Critical Control Point (HACCP) compliance, and facility sanitation.

As might be expected, food costs at the museum's café were higher than targeted (39 percent versus 35 percent). Because the institution's administrators were aware of this situation, they had recently changed from performing physical inventories and inventory adjustments annually to performing them monthly (as is normal practice in the commercial restaurant industry). It should be noted that this higher-than-budgeted food cost had resulted in part from prior efforts to improve the café's food quality and variety: High-end, premade sandwiches purchased from a local restaurant had been added to the menu (and had been sold at prices 30 percent too low), a "chef" had been hired and had acted on a natural propensity to use more expensive ingredients, and a new prime vendor had been retained, raising the possibility that food-purchasing costs on an item-by-item basis had risen.

WHEN GROWTH IS NOT POSITIVE

Offsetting (though by less than half) the increase in food costs was a growth in the café's annual catering revenue, from $70,000 to $120,000. However, even this positive development was mitigated by the way catering selling prices were being set by the special-events department—the all-inclusive terms offered to clients did not allow food costs to be factored in at the desired 32 to 34 percent level, but rather were supposed to ensure the department a profit of 33 percent after all expenses were accounted for. The special-events department did not prepare a budget detailing all anticipated expenses before an event, but rather only estimated expenses based on past practices.

Catering revenue might well have been higher (though costs no lower) if this museum's special-events department, which managed all on-site catered events, had been willing to allow the museum café to cater events and hire its own staff to work catered functions. By using outside caterers for larger events and allowing corporate sponsors to bring in their own caterers, this institution was allowing more than $110,000, or about 60 percent of its total annual catering revenue, to go to outside suppliers instead of keeping it in-house.

While gross sales of alcoholic beverages were on an upward curve, the cost of these products as a percentage of the special-events department's total sales was considerably above the industry norm: They ranged between 31 and 51 percent, where most similar cultural institutions and caterers would typically keep the cost of alcoholic beverages to 20 to 30 percent or less of total sales. In this case, drink prices for guests at catered events may have been set below commercial market levels to encourage patronage of this museum as a special-events destination.

Given the diversity of the challenges facing those administering and providing hospitality services at the museum at that time, we developed our recommendations to offer potential short- and long-term solutions. Suggested short-term solutions included a review of the café's agreements with all vendors (particularly the prime supplier), to ensure that appropriate products were actually being purchased at agreed-upon prices, and periodic audits to check that contracted prices and markups were being consistently applied. Spot-checking all deliveries and weighing all items bought by the pound as they were received were advised, as was the establishment of monthly inventory audits and random inspections by the café's "chef," the manager, and museum administrators.

Other recommendations offered to help this museum's foodservice operation bring food costs in line with objectives were the creation of production records (for comparison with point-of-sale reports) and a written portion guide to be followed by production and serving staff alike. These steps would be most effective if coupled to a comprehensive evaluation of all menu choices and their prices to determine an individual selection's food costs within the menu mix (by taking into account the amount of each item sold in comparison to other selections, waste, products sold at less than full price, and other factors).

We strongly advised that the Wednesday and Friday night dinner program, first- and last-hour services, and the baking program all be subjected to analysis to determine actual costs and profit or loss. Focusing on actual versus ideal labor levels, especially during periods when the café's sales and customer counts were lowest, would likely result in a reduction in required labor hours and number of staff on duty at low-demand times, reducing costs, heightening efficiency, and potentially improving customer service.

Final short-term recommendations included switching the café's tableware from china to high-quality disposables, thus lessening staffing requirements and operating costs. Administrators at this museum were also urged to introduce comment cards in their café, conduct surveys of catering customers to gain feedback on events, and increase the selection of alcoholic beverages to improve guest satisfaction.

REASONS TO CONSIDER AN OUTSOURCED FOODSERVICE

Our long-term suggestions for the museum's restaurant and special-events programs dealt first with the need to address the issue of the café's management. In this instance, based on the factors previously identified, we advised administrators here to consider contracting its restaurant operations either to a professional food management company or a commercial caterer.

It is widely acknowledged that self-operated restaurants in small cultural institutions (under 500,000 annual visitors) usually struggle to achieve optimal top-line revenue and bottom-line profit performance. There are many reasons for this, including the lack of a career path for most foodservice professionals employed by cultural institutions; noncompetitive restaurant management and employee wage scales; an inability to refine and upgrade an operation's offerings, facility, and marketing often enough to stay abreast of the commercial sector; and the complex and sometimes contradictory missions cultural institutions devise for their hos-pitality services programs. All but this last point suggest that administrators must weigh the strengths and weaknesses of any self-operated restaurant or catering service against the potential benefits—and drawbacks—of switching to an outsourced food-service provider. (For more on determining whether to hire a foodservice contractor, the best ways to do so, and how to manage a subsequent relationship, see Chapters 2 and 5.)

At this cultural institution, we felt that hiring an outside firm to operate the café would bring in more experienced personnel who would have access to market-proven resources and systems, helping to enhance the performance and image of the facility with museum visitors and staff. This would logically result in increased check averages and customer satisfaction with the overall museum experience, as well as lower food costs (due to the operator's greater purchasing power) and more-professional management.

It was also advocated that all food and beverage services at this institution be considered and managed as a single entity. Given the gross total amount of annual food-and-beverage revenue being generated at this museum, aggregating all hospitality services under a single contractor or caterer would allow the institution to shed administrative operating expenses and offset staff meal subsidies. It would also bring the museum a percentage of foodservice sales large enough to create at least a break-even situation, rather than the annual loss the café was then incurring.

After two additional years of self-operation, this museum outsourced its foodservice operations. The results in this case have been improved financial results and increased foodservice customer satisfaction.

The possibility was also raised of converting part or all of the underutilized staff dining room into an additional catering space and ending the museum's policy of limiting the right to host special events just to corporate sponsors or the institution itself.

CREATING A STRATEGY TO PRODUCE A SURPLUS

When we provided consulting services recently at a natural history museum, we found administrators' goals for their self-operated restaurants and catering to be similar to those of their peers at other institutions, with the important exceptions of having hospitality services generate an annual surplus and be further developed according to a multiyear strategic plan. This plan was intended to align foodservices more closely with the museum's mission by providing an innovative, high-quality experience that captured the interest of the community.

At the time of our study, this cultural institution operated a full-service cafeteria, a quick-service walk-up deli, and a coffee cart. The museum hosted about 600 internally sponsored special events and some 300 outside-sponsored events annually. Until shortly before our involvement, this museum's restaurant and catering programs had been under the direction of an experienced and capable self-operator manager. Unfortunately for the museum, this manager had recently left, citing limited growth opportunities and a compensation package not commensurate with bottom-line performance. Therefore, part of our charge was determining the most appropriate type of management for the restaurants and catering operation, while evaluating the performance of the current interim self-operator, the department's executive chef.

It soon became clear that one of the foodservice facilities' greatest drawbacks at this museum was the fact that equipment had received no significant upgrading since operations had commenced over a decade before. Menus, too, had gone several years since being updated or even reviewed. Customer preferences were unknown, since comment cards had never been used, but museum administrators we surveyed indicated that their lukewarm appreciation of the foodservice operation would be stronger if menu items were more healthful, freshly prepared, and of consistent high quality; if service staff were friendlier, more efficient, and knowledgeable about the institution; and if eating areas were brighter and more contemporary.

Senior staff here expressed greater satisfaction with current catering efforts but still felt that more consistently prompt meal delivery (especially to breakfast meetings) and the addition of trendier food choices would help to bring this service program fully up to expectations.

Perhaps most potentially problematic for this institution's foodservice, however, was the entrepreneurial manner in which it had been operated. Many formalized procedures and systems commonly employed in the foodservice industry were absent, leaving the museum particularly vulnerable to turnover in key staff. This was acutely exemplified by the recent departure of the talented but unorthodox former manager, which had placed considerable strain on remaining department members, in great part because they inherited very little documentation of the operation's practices. What's more, the interim manager—the executive chef—was making food purchases without specifications for products and failing to keep any production records. It should be emphasized that at most medium-sized and larger cultural institutions (with over 500,000 annual visitors), executive chefs are normally involved *only* in kitchen operations. Having a chef involved in other management-related areas can cause substandard performance and is an indication of poor management structure.

TRACKING THE PROBLEMS

Though there was very little in the way of formal procedures, financial audits, and records of restaurant customer counts and check averages, the former manager had tracked what he referred to as "net sales per person." While this was a useful (though not industry-recognized) measurement, the most recent figures revealed that the more than 1.7 million annual visitors to this cultural venue spent some $1.224 million in its restaurants. This equated to an average spending per museum visitor of only $0.72, compared to check averages of $1.25 to $2.00 typically recorded by visitors at similar institutions.

Further shortfalls in these foodservice operations included an absence of specifications for purchased foods and supplies, vendor contracts, vendor bidding, and price auditing. And although some

30 percent of all food purchases were meat, poultry, or fish, products that are typically weighed upon receipt and compared with purchase orders, product checking and weighing were not part of operational controls at this museum.

Additional areas of concern included long lines at the deli (caused by the facility's cramped layout) and menu variety and complexity. While the scatter-system layout of the full-service cafeteria's servery was commodious and offered the potential for further expansion and promotion of new serving concepts or stations, its decor was tired-looking and gave the impression of not being entirely clean. What's more, as mentioned, the cafeteria's kitchen equipment was older than average in the industry and, thus, more susceptible to breakdowns and subpar performance. On examination, the equipment, work surfaces, compressor cages, and storerooms all evidenced a need for better cleaning practices, and stored perishable items were not being dated, rotated, or held according to industry-accepted practices.

Interestingly, this institution had been conducting annual customer satisfaction surveys prior to our involvement. Although its restaurants scored satisfactorily in areas such as food variety, service, and appearance and were drawing a commendable 28 percent of total museum visitors to their locations, scores were beginning to trend downward. We took this as an indication that there was significant upside potential for the museum's restaurants (it should be noted that visitors at most cultural institutions do not have high expectations for their dining experiences).

A concurrent examination of the institution's catering program revealed that most events were small and that rates charged (determined by the former manager's informal "retail less 20 percent" pricing formula) were likely not covering all actual costs, especially for internal catering. In addition, all catered meals were being completed in the kitchen and transported to catering venues in hot boxes; more professionally managed catering services now usually involves catered meals being prepped in a kitchen and then finished and plated at points of service. It was also found that all catering paperwork was still done manually and that there were no written procedures for booking, billing, or overall program operation.

A RANGE OF REMEDIAL OPTIONS

Three options were identified for the resolution of the management situation:

1. Hiring another self-operator manager
2. Hiring a contract company manager backed by corporate resources, but not outsourcing the department
3. Outsourcing the entire department

For this museum, we endorsed options one and two, reasoning that a contract company manager would introduce missing formal controls and procedures without requiring an investment from the museum. If retaining complete self-operation was preferred, the key to our recommendation was that administrators had to put together a highly competitive compensation package, including bonuses for exceptional performance and a capital improvement budget, in order to attract a top manager still early in his or her career. Even with these inducements, it was estimated that an outstanding foodservice manager would probably be looking for new career opportunities in five years or less, as the only way up for a self-operator of hospitality services in cultural institutions is out. That was why we attributed several potential advantages to hiring a contract company manager, who would incur a lower net cost, would likely arrive well trained, and would be more easily replaced if found unsuitable. Finally, it was important that, regardless of the type of manager to be hired, staff training in customer service be initiated immediately.

We recommended that once the management decision was taken, new accounting procedures should be established for foodservices, including tracking staff meal discounts and internal catering charges and costs, and recording supplies by appropriate product category (i.e., paper, janitorial, etc.). Surprise and scheduled audits of cash handling, accounts receivable, inventory, and purchasing practices were also advised, as were identifying customer counts and current check averages and recording them over time to create a history of demand. It was also advised that it would be good idea to purchase software to manage all office functions. Additional financial recommendations included bidding purchasing contracts annually or biannually, installing a scale to measure upon receipt all products purchased by weight, and drafting written production records.

To begin an improvement program for on-site restaurants, we recommended that both facilities introduce new, more healthful, and contemporary menus and more market-competitive pricing. We suggested that the staff receive new uniforms and that the facility undergo extensive cleaning and freshening. In addition, the museum was urged to retain an independent food safety and sanitation inspection service to evaluate and monitor the restaurant's sanitation and food-handling practices. Such inspections are vital to the correct management of self-operated museum foodservices, which are typically not subject to regular third-party inspections nor privy to reports of findings when such inspections are conducted. The museum was also urged to join its state restaurant association and the Society for Foodservice Management (SFM), to give managers and chefs a chance to network with peers and learn more about industry issues, trends, and best practices. The addition of customer comment cards and creation of a promotional handout designed to acquaint all visitors with the restaurants were also advised.

To meet administrators' long-term objectives for their restaurants and catering, it was recommended that a foodservice facility or restaurant design firm be brought in to help develop a master plan to guide the evolution of existing facilities. The mitigating effect of ongoing cost control and overhead reductions would likely allow some 40 percent of new foodservice annual revenue to fall to the bottom line, providing sufficient funding for ongoing facility renovations. In our experience, the more complete the renovation of older foodservices, the greater the potential for increased future revenue and profits, better operating efficiency, and lower staffing and utilities costs.

Our analysis of the museum's catering led to recommendations that included standardizing internal catering costs, eliminating existing (unintentional) interdepartmental subsidies, pricing events according to formal expense budgets, and initiating a menu development program to make menu selections comparable to and competitive with those offered by leading commercial caterers. It was further advised that catering storage areas receive new shelving units to facilitate more efficient organization of inventory, and that existing tables and chairs used at events be replaced with furnishings of more modern design. To make internal booking and billing more efficient, the museum was encouraged to put these functions on computer and dedicate a staff member to managing the reservation book.

Overall, this museum was urged to develop a business and marketing plan for special events that focused on increasing ties with corporations, sponsoring organizations, and other user groups; set out the marketing mission and goals; and provided a budget sufficient to purchase all needed new materials (such as brochures and menus). We also recommended that this institution advertise its foodservices in member publications and on its web site, improve promotional signage in and around the facility, and create visitor "handouts" that touted dining services. The combined effect of all these enhancements, we concluded, would likely create a more stable and accountable management structure, improve financial performance, increase visitor satisfaction, and tie hospitality activities more closely to the museum's institutional objectives and aspirations.

IMPROVING HOSPITALITY SERVICES TO GAIN A COMPETITIVE ADVANTAGE

Along with keeping foodservices self-operated, the chief goals for visitor-service administrators at one medium-sized art museum were to use their restaurants and catering services to set their institution apart from local peers and other cultural and entertainment venues, while assisting in the raising of new development funding.

We found that the layout of this museum's restaurant, kitchen, catering, and support areas was problematic, with the main dining room and other guest spaces situated on the first floor and cooking and storage facilities located on a different floor. This not uncommon separation of foodservice spaces (also found at the Cleveland Museum of Art, the National Gallery of Art, the Art Institute in Chicago and the Los Angeles County Museum of Art, among others) inevitably engenders additional labor, time, and frustration. Therefore, operations of this design should be targeted for reconfiguration whenever feasible; however, this was not possible in the short term at this museum.

Despite the layout disadvantage, the restaurant and catering department's annual results from two years before our arrival had included a $110,000 operational profit. This was due in part to the skills of a veteran self-operator with a chef's background, an understanding of the museum's mission, and a commitment to continuing professional education. However, the department's most recent annual results had shown a loss of nearly $30,000, primarily as a result of using outside caterers during recent years. Museum trustees and administrators had responded by seeking ways to eliminate this deficit and restore the restaurant and catering department to (sustainable) profitability.

Deep cuts in this department's restaurant staff (from 22 full-timers to 12) was one remedial action undertaken. However, the labor cutback had placed new stress on the department's manager, as she now felt she did not have sufficient staff to handle some catered events, especially larger ones or those scheduled on a Monday, when the museum was closed to the public. Without enough employees to stagger shifts, this manager saw her staff overtime increasing, which, in turn, made her labor costs too high to be affordable for internal special events. This entire situation was being exacerbated by a policy allowing outside caterers to bid for museum-based events and an extremely tight local labor market, reducing the manager's ability to secure enough skilled "on-call" event staff, which would have allowed her to provide catering on a more cost-effective basis. In addition, foodservice was losing workers to the museum itself, as staff were being attracted by the higher wages and less strenuous duties of other museum positions.

A further complication was the decision of senior administrators here to begin charging a $5 admission fee to nonmember visitors coming to the museum solely for lunch. Previously such guests had been admitted free. The restaurant/catering manager estimated that this new policy would cut her customer counts by at least 5 percent, leading to a shortfall in annual revenue of approximately 4 percent (on top of the previous year's significant loss).

On the brighter side, thanks to a convenient location, expansive guest spaces, a pleasing ambiance, and a well-regarded menu, this museum's restaurant was then recording average per capita revenue (income per visitor) of $2.32, a figure notably higher than those recorded by many museums with similar traffic levels. Nonetheless, the manager of the institution's self-operated restaurant and catering programs knew that morale in her department was not high, and she had even developed doubts about her own future at the institution.

UNCOVERING THE REASONS FOR PAST LOSSES

The 200-seat restaurant featured a simple yet sufficiently varied menu, with items such as salads, soups, and entrees being served by museum volunteers. While these servers were friendly and knowledgeable about daily menu choices, they were apparently unconcerned about portion control. Also lacking was any mechanism to attract and record customer feedback, and the various discounts offered to different restaurant customer groups were not tracked by the museum's accounting department. It is worth noting that if these discounts (worth approximately $50,000 per year) had been credited to the restaurant and catering department, it would have had end-year results firmly in the black, rather than registering a loss.

Other practices that were less than optimal included a failure to conduct regularly scheduled physical inventories of food items and other consumable supplies. And though the museum paid restaurant workers a compensation package that was generous by commercial restaurant standards, pay and benefits were still less than those typically offered by similar museum foodservice departments, although fringe benefits were costing this institution more than if a contract management firm were running its restaurant.

The museum's catering program was more clearly troubled, having suffered a 50 percent decline in revenue during previous years and seeing its sales trending down. It seemed most likely that the recent cuts in hospitality services staffing had reduced their ability to stage special events, though the willingness of museum administrators to cater all but the largest of its parties in-house remained undiminished.

As noted, purchasing and inventory controls were not in line with accepted industry practices. Without regular inventories (except of alcoholic beverages), the department did not know the value of its stores and could not compute its cost of goods sold (value of beginning inventory plus purchases minus ending inventory comprise this industry-standard measurement). In addition, items bought "by the pound" were not weighed upon receipt, and the department was buying from some 30 vendors, a number that indicated there was considerable room for purchasing consolidation.

Accounting practices were also lacking, as monthly reports failed to include food and labor costs as percentages of overall expenditures, and the exclusion of discounts, refunds, sales tax, and tips had led to skewed bottom-line figures, producing an inaccurate portrait of the department's actual performance. In addition, because labor costs were in fact running higher than budgeted, this museum's catering program was on course only to break even, rather than earning the profits anticipated by administrators.

ENSURING ACCOUNTABILITY

Because accounting omissions were evident and corrections would yield important and accurate information about the performance of the restaurant and catering department, we recommended that discounts and overtime expenses both be included in future budgets. Adding in the cost of goods sold was also advised, as was an analysis of the restaurant's purchasing program. Beyond ensuring that all items were being purchased at agreed-upon vendor prices, this analysis would involve the spot-checking, weighing, and/or counting of all delivered items. Further recommended controls included initiating monthly physical inventories of all received goods, with management spot-checking staff's tallies.

It was determined that kitchen activity would become more efficient if daily production records were kept and reviewed against point-of-sale reports. Other suggested improvements included the creation of a written portion control guide to be used by all production and serving staff, and formal training in portion control procedures. In addition, we advised that all menu items and ingredients be reviewed to determine how they compared with food cost goals.

Looking at staffing costs and levels, we first recommended that all overtime hours/dollars be identified in the monthly statement of activity. As the foodservice manager felt that staffing was overly tight, our potential solutions included studying daily operations to determine if current labor was sufficient for those needs, including internal catering requests, on-site baking, and new inventorying duties. It further seemed advisable that the museum's catering department begin providing services for the large events currently being handled by outside caterers.

To gain insight into how visitors felt about this museum's restaurant and catering programs, we suggested introducing customer comment cards in the restaurant and conducting satisfaction surveys with guests who had attended special events.

Finally, since administrators here were interested in exploring further the financial potential of their restaurant and catering operations, it was suggested that they initiate a limited request-for-proposal process to determine which benefits (if any) an outside foodservice company might offer. One reason for pursuing this option would be to give the self-operating manager (and some key staff) a chance to join an outside company, thus increasing their resources, wages, and benefits and, potentially, decreasing the stress then marring their jobs and performance. Another alternative was to outsource the restaurant and/or the catering service to an outside foodservice supplier while keeping the current manager as a museum employee. This option would allow her to serve as the museum's liaison to the contract company and keep her in charge of quality control and service performance, but relieve her of the burden of daily operational and administrative responsibilities.

Performing Evaluations and Assessments: Challenge-Solution Case Studies

Despite the busy pace of daily foodservice operations, every cultural institution's hospitality services department should acknowledge the periodic need for a fresh set of expert eyes to review restaurant facilities and catering services. Regardless of whether hospitality services are self-operated or outsourced, the reasons for initiating evaluations and assessments are numerous. Very often a change in senior or administrative staff may herald a new guest services mission or a reexamination of the existing one. Personnel turnover or current managers' failure to employ best practices can also cause formerly profitable or self-supporting restaurants or catering services to lapse into losses that on-site personnel may not be able to reverse. The responsibility of planning new dining or special-events facilities or renovating current ones is often best undertaken with the help of a qualified consultant's analytical abilities, objectivity, and experience.

Yet, whatever the specific motivations behind a cultural institution's decision to engage an outside foodservice consultant to evaluate its hospitality programs, there should be one overriding goal:

Assessments of foodservices should result in recommendations that help align products and services more closely with guests' desires and expectations. When this occurs, visitors will likely stay longer among the exhibits and installations and experience a higher degree of satisfaction overall, while local corporations, civic groups, and private individuals will be encouraged to choose these cultural institutions as special-events venues and recipients of donations and bequests.

Evaluations of hospitality services can also be employed to fine-tune existing restaurants and catering operations, as well as to help guide more comprehensive program changes. Menus can be analyzed, for example, to ascertain how well they reflect contemporary diners' and visitors' tastes. In addition, service styles, marketing and promotional efforts, back-of-the-house management practices, financial systems, and assorted operational checks and balances can be reviewed individually or in any combination. The presentation of a report at the conclusion of such an assessment can help senior administrators determine not only what steps need to be taken to ensure that their foodservices perform up to acknowledged industry standards, but also where and how savings can be realized, how and where to expend precious capital funds, and how best to position their institution as the local destination of choice for cultural education and entertainment.

PLANNING TO MEET GUESTS' EXPECTATIONS

We were recently called in to a midsized museum of art to help administrators determine the most effective ways for their dining services to reach certain visitor-focused objectives. These goals included having foodservice provide products and services that met the expectations of the greatest possible percentage of visitors, particularly children and teens; offering special and catered events of sufficient quality to create memorable experiences for current and potential museum supporters; providing reasonably priced internal catering services; increasing revenue from special events; and integrating the museum café's menu and decor with selected art exhibits.

Evaluation of the steps necessary to attain these objectives began with the identification of several constraints. Space was at a premium in this institution, meaning that foodservice areas could be expanded only at the expense of such core functions as education or exhibition displays. In addition, food and beverages could be introduced into gallery spaces only if the integrity and safety of the museum's collections could be ensured. It was also noted that an expansion of special-events functions would likely result in greater wear and tear on the museum's physical plant and put further strain on such key personnel as housekeeping and security staff, who already seemed hard pressed to keep up with the existing schedule of catered events.

On-site restaurant operations were reviewed next. Operated for years by a local off-premise catering company, the museum's café was located below ground level, adjacent to a secondary entrance. Depending upon the configuration of furnishings, this café could accommodate between 70 and 90 guests at any one time (though it was then attracting barely 80 customers a day). It had a table-served menu and a self-service salad and soup buffet; beer and wine were served, and a children's menu was available. The café's kitchen was equipped with a refrigerator, tables to hold hot food cabinets and sandwich make-up tables, microwave ovens, and a convection tabletop oven, allowing menu choices to be plated and served as ordered. However, all foods then being served at this café were being produced off-site by the catering firm contracted to run the facility. Cooking and serving ware were cleaned in a small dishwashing machine and a three-compartment pot sink, while a servers' station supported beverage service.

According to data provided by the off-premise caterer (which had been in place for nearly a decade), the museum café's average per capita revenue (total café revenue divided by total museum attendance) had fallen from $0.67 to $0.31 during a four-year period, with losses running as high as $0.22 per visitor (though losses had begun to abate at the time of our evaluation). While such food-service losses are not uncommon for museums with annual visitor counts of between 200,000 and 250,000, and are particularly prone to occur when a restaurant offers table service rather than self-service, this café's per capita revenue, even at its highest, was less than half that typically recorded by comparable foodservices in comparable institutions.

As for catering, this museum was hosting an average of 21 special events each month: 12 internal functions and 9 external affairs. (In addition, an average of 10 events and over 50 meetings were being staged every month that required no catering support but did demand the involvement of other support staff, such as housekeeping and security.) Catering support came from a small main-floor service kitchen and liquor storage rooms. Unfortunately, the service kitchen was not conveniently located in relation to the several main-floor gallery spaces that regularly served as special-events venues, forcing catering staff to plate meals and set up for service in the museum's corridors. An ongoing, separate concern on the part of this institution's curatorial staff and administrators was the fear that foodservice activities, such as transportation of meals in sharp-edged, steel-cased, hot holding cabinets, could lead to damage to the exhibited art, even though the works were enclosed in protective cases. Nonetheless, this museum supported special-events activity, believing it a key to growth and community service, and promoted its special-events capabilities via advertising in the local business press, contacts with the visitors bureau, hosting chapter meetings of the International Special Events Society, and distributing a marketing brochure. It had even recently begun placing ads in bridal magazines and sending representatives to bridal fairs to gain further recognition as an appropriate setting for weddings.

The catering firm that managed both the café and all catered events (other than a few sponsored by corporations, which were given the option of using their own caterers) seemed to enjoy a generally positive reputation among this museum's administrators and staff. This regard was founded on three factors: high-quality food and service, flexibility and cooperation in making arrangements, and sensitivity to the museum's concerns about protecting its collection. This last point caused several administrators at this institution to tell us that the current caterer's ability and willingness to align itself with the museum's culture made them reluctant even to consider calling in and "training" any other outside caterers. The current arrangement held apparent advantages for the incumbent caterer, too, allowing this firm to more than offset four-figure losses at the café with annual profits of between $50,000 and $60,000 on roughly $250,000 a year in catering food and beverage revenue. When we looked at the overall foodservice financial picture at this museum, we saw that more than 33 percent of all revenue came from liquor sales at special events (this museum self-operated its alcohol beverage services), and that liquor sales and facilities rental fees together constituted more than 57 percent of total average annual income to the museum. It thus seemed likely that more revenue *and* a more balanced income stream could be created by raising event and facility rental fees and by charging extra for housekeeping and security services, rather than including them in the base fee, as was the practice then, provided, however, that such fees and charges continued to be competitive with other local area special event venues.

IDENTIFYING SHORTCOMINGS

With an initial review of operational components now complete, it was time to see which issues were most directly affecting this museum's ability to reach its foodservice objectives. The café's low per capita sales indicated that it was serving a relatively small percentage of visitors and staff. Contributing to this situation were the facts that some 30 percent of all total visitors to this museum were children (who tend to consume less than adults); the café's menu, decor, and service style, which all appealed primarily to mature adults rather than to families with youngsters; the dining facility's below-ground location, away from the ground-floor galleries and the museum's main entrance, which caused it to be out of sight and thus out of mind for many visitors; and the lack of signage inside the main building directing visitors to the restaurant.

One way that most of these shortcomings could be eliminated would be to relocate the café to a more heavily trafficked area. Various ground-floor gallery spaces presented attractive alternative locations, especially one that was contiguous to a sculpture garden and thus afforded opportunities for outdoor dining.

Another remedial step would be to increase the facility's capacity to support desired future growth. Current capacity was then limited both by the number of seats (±70) and by extended guest waiting times caused by the café's fine-dining-style table service. We found, for example, that during three-hour daily lunch times, each party of guests was at table for between 45 and 60 minutes. It still seemed reasonable to assume, however, that daily customer count could be increased to around 120, raising the café's "capture" rate to between 16 and 18 percent of total daily visitors.

A third obstacle to enhancing this museum's restaurant program was the practice of accepting meals produced off-site by the commercial caterer and preparing them in the on-site finishing kitchen. Without the support of a museum-based production kitchen, entrée menu choices at the café were necessarily limited to those that could be produced elsewhere, maintained safely during transport, and held further on a steam table prior to service—all without losing quality and attractiveness.

This assessment resulted in a list of specific recommendations. These included relocating the café closer to main visitor traffic areas, changing its serving concept from table wait-service to buffet (self-) service, revising menus to include more choices attractive to younger people, adding "modest" final-preparation capacity to the existing finishing kitchen, and determining the ideal location for the café via a facility-use study designed to guide the optimization of all the institution's programmatic needs.

ASSESSMENT IDENTIFIES IMPROVEMENT OPTIONS

Based on all the evaluated factors, the decision of exactly where to relocate the museum's restaurant required administrators to consider a variety of options. As mentioned, several ground-floor galleries offered suitable locations, but situating a dining facility amidst exhibition halls held the drawbacks of reducing available exhibit areas, imposing product delivery and trash removal requirements, and requiring the formulation of a diner/visitor admission policy. Other options included moving the café to a major trafficway between a parking garage and the museum itself, and inviting a local restaurateur to develop a replacement fine-dining restaurant on the roof of the museum's main building. Because so much of the museum's operations could be affected by the eventual disposition of the café, it was strongly urged that a formal master planning process for the development of foodservice for the entire museum facility be initiated. Until such time as that process produced a comprehensive plan, it was recommended that a foodservice cart stocked with popular hot and cold beverages and snacks (including as many nationally branded products or same-quality equivalents as feasible) should be set up in the museum's Great Hall on the ground level. This cart would be supported by the addition of bistro-style tables and stacking chairs in this exhibit area. As for the café itself, we noted that introducing signage promoting the restaurant's presence and theming its menu and decor to accord with major museum exhibits would likely result in higher visitor counts, regardless of style of service.

Regarding this art museum's special-events spaces, a more strategic usage strategy was recommended for the short term, while plans were made to improve liquor storage security, provide more area for the staging of sit-down catered meals, and select a single space (of approximately 3,200 square feet) where as many as 250 guests could be seated. In reference to using special-events spaces more strategically, recommendations included hosting large presentations or speaker-focused events in the museum's auditorium rather than the Great Hall and restricting sit-down meal service to areas of the building where damage to the art and physical plant could be minimized. Other ideas for limiting the possibility of damage were to fit clear acrylic corner protectors on corners in areas where foodservice equipment or furnishings were to be transported or set up, and using *stackable* tables and chairs that could be most securely transported by dollies or hand trucks for all special events, thus minimizing chances of contacting corridor walls or exhibit display cases.

IMPROVING CATERING CONTRACTS AND TERMS

We found this museum's arrangement with its outside foodservice provider to be generally fair and appropriate, especially as the operator was able to use its virtual monopoly on catering to offset losses incurred in running the café. We advised that a new three-year contract with the provider should be negotiated to cover the café and catering services. This agreement would preferably be made renewable on an annual basis and be subject to cancellation on 60 days' notice, with both options held at both parties' discretion.

We further urged that the existing outside foodservice provider be granted exclusive catering privileges under a profit-and-loss contract, with only a few annual corporate-sponsored events exempted from this agreement, and that all catering menus and prices for external and internal events be specified in the contract (with some flexibility for custom catering), as well as the conditions under which prices and/or menus could be changed. The need to offer exclusivity to the museum's foodservice operator arose from the relatively small food and beverage volume then being generated by this museum, the ongoing need to offset café losses, and administrators' desire to minimize the time spent training other caterers' personnel to avoid the risk of damage to the museum's collection. However, it was also advised that if negotiations to implement these revised terms failed, it would be best for this museum to initiate a formal foodservice RFP process, including interviewing several other foodservice operators to determine their interest level, and begin the process of selecting a new operating company.

The current discount on internal events being offered by the provider (15 percent) was particularly modest (due, again, to this institution's self-operation of its liquor service) in light of accepted industry practices that typically call for discounts of 20 to 30 percent for internal catering, depending on the overall volume of foodservice revenue and location profitability. While negotiating an increase in this discount rate appeared feasible, it also seemed likely that the caterer would ask the museum to offer contractual concessions in other areas.

Another concern was the need to revise current facility-use contracts between the museum and outside groups and organizations having events at the museum so that policies and procedures for special events were specified in sufficient detail. If this was done, the museum's special-events staff would no longer have to expend significant time and effort explaining policies, modifying guests' expectations, or striving to elicit compliance with tacit under-standings. It also seemed useful for administrators at this museum to begin compiling a special-events handbook or similar documentation describing clearly the policies and procedures relevant to conducting a successful event at their facility. This handbook could be incorporated into use agreements and also be used as a reference guide for staging internally sponsored events.

Once these administrative and contractual issues were resolved, it was recommended that the museum next seek several operational enhancements from its outside foodservice provider. These included improvements in food quality, operator participation in visitor/customer assessments, and greater involvement in marketing the museum as a hospitality services venue to its clients.

Our evaluation to this point had indicated that the outside food-service provider was widely considered capable of offering catered food and beverage services of a quality consistent with the museum's objective of positioning itself as the premier cultural venue within its community. Nonetheless, improvements were still needed to eliminate occasional subpar items such as wilted salads and stale rolls, which indicated less-than-optimal quality control at the catering firm's commissary. To minimize quality shortfalls, this museum's administrators were advised to ask the foodservice provider to make whatever changes necessary in purchasing, storage, handling, and delivery practices to ensure that only first-quality products were provided. Should these remedies prove unsatisfactory, the museum would best be served if its provider was required to seek outside expert advice and/or a partnership with a firm or individual with experience in operating on-site restaurants in a museum locale.

It also seemed appropriate to recommend ending the practice of assessing visitor satisfaction with menu choices in a trial-and-error fashion, and to initiate systematic surveys of guests' and visitors' foodservice preferences and expectations, with the foodservice provider's participation.

Finally, to allow this museum to receive maximum advantage from the privileges extended to its foodservice provider, it should expect that firm to develop and present a plan for promoting the institution as a special-events venue and to guarantee a minimum number of events to be held on the premises annually.

WHY DEVELOP A BUSINESS PLAN?

With enhancement objectives for catering established, our evaluation next indicated that a business plan including financial projections, assumptions, and budget needed to be developed for this museum's special-events program. As would be the case at any comparable cultural institution, this business plan should begin by stating the qualities that distinguish this museum from other local cultural education and entertainment venues. The plan would also need to determine the primary markets (audiences or visitor groups) for the museum's special events, a competitive pricing strategy, a promotional program (including advertising) that would identify the activities necessary to increase special-events business, and the amount of human resources needed at each growth threshold, including personnel required for special events, housekeeping, and security.

The recommendation to devise such a business plan was perhaps our most important assessment, for two reasons. Without a plan of this type, museum senior administrators would be unlikely to invest the resources necessary to achieve identified objectives. Also, once a hospitality services business plan was approved at the highest institutional levels, restaurant and special-events administrators were likely to gain new legitimacy and status within the museum's organization.

PROVIDING FOR OPTIMAL PROGRAM QUALITY

Providing foodservice program evaluations and assessments at a very large museum is inevitably a complex task, especially when on-premise hospitality services are expected to draw visitors with products and services of equivalent quality to that of the museum's world-famous exhibits. At a second, larger art museum, our first task was to identify administrators' goals and objectives for their foodservice operations. These included the provision of quality menu items at prices deemed reasonable by the public and staff, earning sufficient revenue to allow the museum and foodservice operator to at least cover costs and preferably allow both parties to make a profit, upgrading food presentations and expanding menu variety to appeal to a broader cross section of visitors, enhancing customer service so that it was consistently prompt and courteous, and positioning the institution's public restaurants as desirable dining destinations rather than just on-site conveniences.

At the time these goals were being formulated, all restaurant operations and catering services at this museum were being managed by a local contract foodservice firm that had served the institution continuously over the course of many years. The contractor's catering capabilities, however, were limited; an outside catering firm was being employed to handle "fancier" high-budget events. The overall foodservice lineup at that time included staff dining, public restaurants, as well as vending machine and catering services.

ASSESSING STAFF FOODSERVICE

One of the staff dining facilities, which provided both light morning snacks and beverages as well as full luncheon service, also had the distinction of being used by the museum's administrators for entertaining donors and VIPs, in addition to hosting general staff dining. This facility offered seating for 120 (a large percentage at common tables) and was receiving prepared products from the institution's main kitchen, despite the fact it had its own, fully equipped production kitchen. Despite its dedication as a staff dining environment and a 30 percent discount on all purchases made by museum employees, most personnel queried indicated that they did not patronize the facility, citing its distance from work areas, among other reasons. A study of the contractor's customer counts revealed that during the past year only about 20,000 staff members, including visitors and guests (an average of about 80 a day), had been served. Given that this museum had hundreds of staff on the premises, this facility appeared to be underused by its potential customer group.

Naturally enough, the contractor had sustained an operating loss for this particular facility during the previous year. Observation also revealed that morning foodservice here utilized a cash box, not a cash register, making determination of morning sales per customer difficult and most likely slightly distorting the facility's overall check average. The lunch menu still relied heavily on traditional meat entrees, which were less healthful than the employees wished and which lacked an air of excitement; worse yet, salad items and dressings presented on a buffet were not being held in a chilled state, creating a potential health hazard for customers and the museum contractor alike. It should be noted that the contractor here had previously suggested closing this facility as a staff restaurant, reserving it only for catered events and eliminating the need for the museum's annual subsidy. Another option for profitability was to maintain this facility's restaurant services but to freshen the menu and adjust item prices sufficiently upward so that diners would be paying a larger share of operating costs.

Another staff dining facility, a 75-seat cafeteria, was used only by museum employees and their guests. Menu prices here, too, were set at levels 30 percent below those paid by visitors in the public restaurants; however, not all menu prices were posted, and some items displayed pricing inconsistent with the established 30 percent discount. Also noted was an absence of sneeze guards at the salad bar and a lack of ice or other chilling elements where greens were displayed, as well as inconsistent use of protective gloves by kitchen and serving staff. Based on the contractor's records, this cafeteria was drawing about 35,000 customers annually, or roughly 140 average per day, and the average check was approximately $2.50. An immediate assessment offered to this museum's administrators was that the cafeteria could be converted to a by-the-ounce concept (menu items sold by weight), supported by a make-your-own-sandwich option. This would allow the elimination of one serving position, and, if prices were properly positioned, customers would be paying for much more of the actual cost of their meals. This shift in serving concept seemed most advisable, since this cafeteria was attracting a very small percentage of available museum staff on a daily basis.

ASSESSING PUBLIC FOODSERVICE

We next turned our attention to evaluation of this museum's public foodservice facilities. One, a café with seating for about 100, was located on the museum's second floor and served lunch only. The café had minimal kitchen resources and so received the vast proportion of its menu selections from the main cafeteria kitchen. Service style here was tableside, with menu items and themes being revised four or more times a year in response to changing exhibits and seasons. According to the contractor, this café had served some 43,000 guests in the most recent prior year, or an average of about 120 customers a day; the average check was about $9 per person.

While these numbers are only a little below average for similar types of museum-based restaurants, the contractor also pointed out that the facility had numerous slow periods during the year (depending on the extent of exhibit activity) and recorded about one-third fewer sales during the week than on weekends, when receipts were averaging approximately $1,800 per day. Since the contractor claimed that the break-even point for this café was about $2,200 per day, it had, not surprisingly, recommended opening the restaurant only on weekends. Although it is never ideal to close a cultural institution's public restaurant during business hours, it is not uncommon for institutions with multiple foodservice facilities to shut part or all of one or more facilities during off-peak traffic hours or when visitor counts are lowest during the year. A similar practice is followed by most public restaurants, theme parks, and food courts, where seating sections and/or serving stations are opened and closed as demand dictates. That said, it did appear that the contractor and museum were doing everything possible to optimize operations relative to daily customer counts at this location.

The second facility, a ground-level café also seating about 100, was located in a high-traffic, high-visibility area of the museum. It, too, was recording average checks of approximately $9, but its annual customer count of 70,000 translated into an average of about 195 daily guests, a little less than one-half that of its upper-story sister operation, which was situated in a less busy, less visible location. As with the upper-floor café, the main cafeteria kitchen was providing prepared foods for meals presented here, with culinary support coming from a small kitchen and converted museum storage spaces, a setup that served to limit menu variety.

Another of the public facilities was a very large cafeteria, whose spacious servery seemed well able to handle the 320,000 annual customers (about 900 average per day) reported by the contractor.

In addition to a very large production kitchen, this cafeteria's back-of-the-house also contained a bake shop, refrigeration, support storage, and offices. However, the cafeteria offered menu selections that most museum staff perceived as "boring," "institutional" or, at best, "okay." We also found a less-than-fastidious attitude toward the facility's cleanliness and upkeep. For example, both the servery ceiling and its fixtures appeared to be in need of cleaning, wallpaper covering the front fascias of serving counters was dirty and beginning to peel, and doors connecting the kitchen and servery also showed signs of wear and neglect. In addition, point-of-sale signage was inconsistent in appearance and contents, with some stations' signs displaying all available items and prices and others showing less-complete information, while pricing itself was inconsistent for the same products at different service points. Yet despite its most evident shortcomings, the cafeteria was recording an average check of $6.25, which was higher than in most comparable institutions (at the time of this evaluation several years ago)—and which led some museum staffers to comment that prices here were too high for typical visitors, especially families.

With this evaluation in mind, we identified several areas where operational savings could be attained. One that immediately stood out during our assessment was the approximately $65,000 the contractor had spent during the year prior to our review on paper napkins imprinted with the name of this museum. We have found that there is little or no return on the use of imprinted napkins (in fact, they typically become "free" visitor souvenirs, greatly increasing consumption), and their elimination can offer significant savings.

A second attainable savings area was identified in the way cafeteria tables were being bussed. It was noted that if staff bussing was eliminated in favor of self-bussing by customers, not only would the contractor be able to reduce facility labor costs, but a greater number of clean tables would likely become available for arriving customers more quickly than when staff cleared meal debris, especially during peak traffic periods. To make this change effectively, however, equipment and traffic flow alterations would have to be instituted and customers would have to be informed of their new responsibilities via informational table tents.

Though this cafeteria was considered a public facility, the operating contractor had determined it to be desirable to serve full breakfasts to museum staff here on a daily basis (with the 30 percent discount). When asked why this breakfast service was not provided in the staff cafeteria, contractor representatives said it was more cost-effective to do so in the public facility. Yet opening this large facility every morning to provide meal service to a comparatively small number of customers was likely incurring costs in areas such as utilities, maintenance, and labor that exceeded the value of those sales. Thus it seemed evident that the preparing and serving of full breakfasts for museum staff members should be reviewed and, if not a continued requirement by the museum, its elimination or relocation considered as a cost-reduction option.

The next public facility at this museum was another café, offering upscale coffee drinks, biscotti, and other light snacks. During the prior year, this facility had served more than 52,000 customers (average about 145 per day) and recorded a $5 average check. While that check average seemed high given the mainstays of the menu here, it may have reflected the sale of recently added upscale sandwiches and salads, which, while attractive, may well have been cannibalizing sales from the nearby cafeteria's own sandwich and salad offerings. The most evident shortcomings noted here were a lack of eye-catching, commercial-style lighted menu boards and signage, and product merchandising and presentations that did not emulate those found in high-end retail coffee or gourmet shops. Since this café occupied a highly trafficked location, it was important that signage and product displays create strong visual impacts on customers while staying within this institution's parameters for understated good taste.

Our examinations next led us to the museum's vending machine and catering programs. At that time, vending machine services consisted of a bank of snack and drink machines located in a remote staff break room. The outside contractor supplied these machines and paid this museum a 15 percent commission on all sales. This revenue, approximately $1,500 per month, was turned over to the museum employees' association.

Our final areas of assessment here included menu development procedures and opportunities for further hospitality services growth. At the public cafés, for instance, menus were scheduled to be changed four times annually. Helping to determine each menu's makeup was a committee of several museum administrators, whose job it was to taste proposed menu items and make recommendations to the foodservice contractor. As expressed during our evaluation, the goal of this committee was to develop menus that would offer guests the experience and tastes of regional American foods, presented in such a manner as to evoke the image of service in a private home.

Other menu selection criteria included product availability (seasonality), cost, practicality of production, ability to be presented consistently, and time required for service. Once item selection was finalized, these products were photographed in a plated state (to assist the contractor's kitchen staff with assembly) and their pricing was negotiated between the museum and the restaurant operator. One problem noted during the assessment of this menu development process was that museum administrators felt that approved products were not consistently being presented to customers as depicted, despite their detailed and extensive input.

IMPACT OF MULTISUPPLIER CATERING

Internal day-to-day catering, as previously noted, was being handled primarily by the contracted foodservice provider. In addition, however, this art museum was often the site of VIP special events—exhibition openings, directors' functions, and others—catered by a group of high-end local off-premise caterers. This multisupplier approach to catering had been adopted because museum staff perceived that their foodservice contractor was too expensive, provided catering foods and services of inconsistent quality, and did not have sufficient on-premise management resources to give the time and attention to detail required to carry off such high-visibility events successfully. (It is common in cultural institutions that either self-operate their foodservices or use a foodservice contractor to hire name-brand, high-quality caterers primarily for external catering and VIP internal catering. This is commonly done to satisfy special-interest donors, sponsors, board members, and the like, though there are exceptions where a self-operated or contractor-operated exclusive catering program works very well.)

The upshot of the division of this museum's catering business was that outside vendors were receiving an annual revenue flow well into seven figures for providing food, beverages, and related support (valet parking, flower procurement and arrangement, linen service, and entertainment). So while the use of high-end local caterers was giving museum administrators assurance that their most deluxe events would be produced to everyone's satisfaction, the practice was also reducing income and profits for both this institution and its contract foodservice partner.

DEVELOPING GROWTH OPPORTUNITIES

When opportunities for the growth of hospitality services here were being regarded, we first noted that the contractor had reported serving only approximately about 10 percent of all visitors to this museum. The desired goal for museum administrators was to raise this figure to a minimum of between 16 percent and 20 percent, while the contractor felt that somewhere between 10 percent and 15 percent was more realistic. Based on this museum's total public restaurant seating and service capacity, 16 to 20 percent was noted to be very conservative.

We advised that market research be conducted with museum staff and visitors to help determine the demographic patterns of restaurant customers, in which operations they had dined and why, which previsit information (if any) influenced their selection of dining venue, and, perhaps most important of all, which visitors were choosing *not* to dine at any facility, why not, and what it would take to gain their participation. Such market research should consist of "intercept surveys," that is, interviews conducted according to established questionnaires, with samples from three groups: all museum visitors, museum staff, and customers at staff and public foodservice facilities. The bottom line for all this information gathering would be the ability to develop restaurant operations comprehensively, so that they best met the needs of both visitors and staff.

The next logical concern to consider was whether administrators needed to review the assignment of responsibilities for the management and administration of the foodservice contract by the museum. Among the areas in which managerial authority appeared to need clarification were menu planning, special events, and day-to-day restaurant operations. Although it was evident that the relationship between this museum and its contractor was reasonably positive, better delineation of which party was responsible for different program aspects might well lead to improved program performance.

The last question posed to the museum's administrators at this time was whether they were willing to use their current contractor more often to prepare and conduct (at least) some of the institution's high-end catered events. If so, sound foodservice administrative practices would dictate that the contractor be held to a set of minimum catering performance standards and that a determination be made as to whether these standards could be consistently met by the incumbent foodservice firm's hourly staff and management.

HANDLING CONTRACT AND POLICY ISSUES

The next recommendation was to advise the incumbent contractor in as timely a manner as possible of the museum's intent to extend the current contract, provided the foodservice management firm was willing to modify this agreement to remove ambiguities in areas such as management responsibility and profit sharing. In addition, the senior staff at this museum were advised to determine the extent of foodservice subsidy by area and department (if any) they wished to continue to pay. The museum was subsidizing some of its dining operations, and administrators were eager to see significant improvements in such areas as restaurant visitor customer counts and gross sales.

More problematic was the issue of staff participation and subsidies, as the museum's leaders had to decide whether their primary concern was to maximize financial returns from their restaurants, leading to a break-even or even a profitable operating scenario, or to continue to offer the staff the benefit of meals priced about 30 percent below market levels. Based on factors particular to this institution's dining program, it was recommended that one of the staff dining facilities be closed for lunch service (though kept open to provide beverages and snacks) to maximize participation at remaining luncheon venues. A corresponding cost-saving recommendation was to reduce the subsidy on staff meals from 30 percent to 20 percent. It is worth emphasizing that subsidized employee meal programs have all but completely disappeared from workplace foodservices, save for the capital improvements, spaces, and utilities required to operate staff dining facilities. Most museums by now have eliminated such dining subsidies as offering staff free meals or paying all or part of foodservice operating costs. Now, at most, they provide staff and volunteers with a 10 to 20 percent discount in public dining operations.

Based on assessments of various restaurant and special-events operations, we strongly urged the museum's administrators to ask the incumbent contractor to seek improvements and report progress within 30 days on a list of identified concerns. These included the expense versus effectiveness of imprinted napkins, the desirability of a switch to self-bussing of restaurant tables, whether to introduce a by-the-ounce program in the staff cafeteria, heightening sanitation standards in the front and back of all foodservice facilities, enhancing less-than-ideal food presentations, upgrading unit signage and eliminating all pricing inconsistencies, aligning menu selections more closely with visitor tastes, and determining the optimal location for staff breakfast service.

For its part, the museum was advised that its staff needed to do a better job of maintaining the physical resources, by cleaning reflective ceilings and light fixtures more regularly, replacing peeling wallpaper, fixing deteriorating plaster in walls, and consistently undertaking other repair and maintenance tasks. When and if new contractual arrangements were worked out with the incumbent foodservice provider or a new contractor, responsibilities for cleaning and maintaining foodservice resources would need to be better delineated.

This led to another key consideration: whether this museum was ready to consider putting all public and staff foodservices out to bid and to make a change in operators if a better overall offer was submitted. It is worth noting that when a client organization, such as a cultural institution, has maintained an ongoing relationship with a single foodservice management provider over an extended period of time, other contract management firms may choose to pass on a chance to bid. This is because potential new suppliers may not be willing to commit the considerable work, time, and dollars required to prepare and submit a qualified bid if they feel the client organization has requested proposals solely as leverage on its current foodservice provider to maintain its current program under more favorable terms.

Another option is to renegotiate the contract with the incumbent contract management firm. (Unless a relationship has soured beyond repair and/or it can be clearly determined through the assessment and evaluation process that an incumbent operator is *not* the best and most up-to-date service provider available, we always recommend a renegotiation approach as the first option.) One alternative, based on the strengths and shortcomings of this museum's hospitality services program, would be to put the incumbent contractor into a probationary state for a predetermined time, during which measurable improvements and goals would have to be attained if the current relationship was to remain intact.

Engaging in a comprehensive RFP process was strongly recommended, since it would provide this museum's administrators with the best way to determine quantitatively how well its restaurant and catering programs could perform in all areas of food and beverage service for guests and staff, as well as special events and vending as compared to other similar large museums that outsource their foodservice operations. It is very important that all RFPs be written in such a way as to ensure an apples-to-apples comparison of various potential providers' replies, so that administrators can determine which foodservice operator best matches their needs and objectives.

CREATING TARGETS FOR IMPROVEMENT

The next site of our evaluations and assessments was at a small museum of art. Here, administrators and museum staff had identified a clear set of goals and objectives relating to the performance of their restaurants and catering program. In priority order, these were: ensuring that the sole full-service restaurant at least broke even and that catering services turned an annual profit; offering food and beverages of the highest quality to museum guests, visitors, and staff; providing services that were consistently customer-focused; providing dining and catering services that were perceived by visitors and guests as a great experience while functioning as integral components of the museum's overall operations; ensuring that food-services offered a diverse array of services; and having restaurant managers and foodservice staff treat all customers (visitors, guests at special events, and museum staff) as equally important.

One of the reasons administrators at this institution recognized a need to assess and set specific goals and objectives for their restaurant and catering program was that the management of hospitality services had changed three times during the past few years, switching from a foodservice contractor to a commercial caterer and then to self-operation of the restaurant and smaller special events. The former off-premise caterer was still being retained to handle larger special events and, since it held an alcoholic beverage license, to provide all alcoholic beverage catering. This caterer was paying the museum a 15 percent commission on all on-premises catering sales, with total annual catering revenue running between $175,000 and $200,000. Yet, despite the significant amount of business this museum was providing, museum administrators did not believe that the commercial caterer was doing enough to promote the institution as a catering venue, probably because it competed with the caterer's other clients for business in the local community. It should also be noted that because this commercial caterer handled only the museum's larger events, to have the caterer promote the institution as a catering destination of choice for smaller events could have resulted in a conflict of interest, with the caterer spending its own money to funnel business to its competitor, the museum's foodservice operation. In addition, this same caterer had been retained to operate a snack bar during museum-sponsored outdoor symphonic performances, though it was unclear whether the institution was receiving a percentage of these sales. More unusually, this catering company was also acting as a consultant (though without a contract) to the museum's restaurant and catering program, charging $25,000 per year for its services.

This spreading of responsibilities for hospitality services also included the marketing and sale of the museum's picnic and tour group services, which were being managed by a member of the institution's public relations department. What all this boiled down to was that food and beverage services were falling under the administration of three organizations: museum public relations, a self-operating management group, and a commercial caterer.

Although the museum restaurant's design and decor were well aligned with the parent institution's image and mission, revenue at this facility was extremely variable, fluctuating from week to week and season to season according to group tour volume, visitor admissions, and even the weather. As a result, customer counts ranged widely, with conditions in the restaurant ranging from overflowingly crowded to all but deserted. Complicating the unpredictable nature of the restaurant's revenue was the fact that it had been declining for the past three years, from $265,000 to $186,000. Closer examination suggested that this operation could be turned around, since it already offered an extremely popular Sunday brunch program. Given appropriate advertising support, high menu quality, good service, and an improved value/price relationship, we believed it could attract improved daily lunch business.

Before this could happen, additional restaurant-related shortcomings had to be identified and strategies developed for their elimination. One easily observable oversight was the absence of restaurant menus posted at the facility's entrance or anywhere else in the museum. Given that this institution was widely considered a tourist destination, the lack of informative and/or promotional materials was putting its restaurant at a significant disadvantage in its efforts to attract one-time visitors. A dearth of common foodservice signage was also noted in the restaurant's food presentation area, where merchandising suffered from a failure to use decor items as attractive props. Perhaps even more significantly, assessment revealed no written policies or procedures governing purchasing, employee training, sanitation, food production, recipes, and other key elements. In addition, the restaurant's management was not holding weekly meetings either with catering managers or internally, leading to less-than-optimal communication between the various foodservice enterprises and a propensity to leave problems unresolved while daily duties were dealt with.

The museum's in-house catering operations were in a more positive business situation at this time, with revenue over the past three years having risen from $295,000 to $442,000. One reason for this growth was the active marketing of catering services to the local business community and the recent initiation of a bridal fair at the museum. Another was the good service provided at special events, which led to generally high scores for catering on customer surveys.

In addition to its restaurant and catering programs, this museum offered snack and beverage items during sponsored outdoor concerts. Our initial evaluation was that these services had significant growth potential if supported by an appropriate plan for pricing, packaging, and promotion. Foodservice here also included a retail baking department, whose sales had held relatively steady during the past three years but whose product merchandising was not up to commercial standards and whose selection was a mix of fresh-baked and frozen-dough-based products, leading to a perception of inconsistent quality.

When the museum's snacks programs and vending and coffee services were all added to the array of foodservice operations, we estimated that the combined activities were generating some $1 million in annual revenue, despite limited promotion and marketing and a restaurant operation that was inadequately customer-driven. Thus there was a promising potential upside for foodservice.

CONTROLLING THE IMPACT OF A MANAGER CONVERSION

Given these findings, our next consideration was determining the future management status of the museum's restaurant and catering operations. Because administrators had indicated that foodservice operation was not a part of the institution's long-term mission, it was logical to consider an outsourcing alternative, provided that local firms could provide effective management, successful promotional programs, and adequate financial returns to the museum.

A conversion from self-operated to outsourcing foodservices is complex and can raise a variety of issues affecting the short-term performance of a program. These issues include the reactions of incumbent managers and chefs, who will almost inevitably begin looking for new jobs when they learn of the pending changes, since an incoming contractor will want to (and should) bring in its own management team. Hourly employees are also likely to feel less secure in their jobs, and though a contractor will usually seek to retain most of these staff on a trial or probationary basis, day-to-day operations may well be disrupted. These problems may emerge well before any transition actually occurs, for in order to attract the contractor best qualified to manage a particular cultural institution's restaurants and catering operations, an RFP document must be distributed to as many potential candidates as possible. Since this should involve advertising the opportunity with the restaurant industry trade press and state restaurant association, current managers and staff will soon learn of the situation, and relations with other institution departments or programs may suffer. The alternative is to distribute RFPs confidentially and to provide private, off-hours facility tours; while this can delay discovery of a pending foodservice change among incumbent personnel, it can also extend the transition process significantly.

In the situation facing this museum, it was recommended that the RFP process be conducted publicly, preferably during the winter (the slowest period for restaurant traffic during the year), and that, if necessary, catering would receive assistance from a local company to ensure that events could be staged during the transition.

Our next recommendation focused on sanitation and cleaning practices, since prior observation had revealed substandard conditions in the restaurant and its kitchen, the bakery, and the snack facility. Besides ensuring that a clean-as-you-work policy was enforced in foodservice operations, contacting the state restaurant association for information about ServSafe seminars and HACCP programs and initiating a relationship with a professional foodservice sanitation company were advised.

Conditions at this institution were such that a thorough upgrade of facility sanitation would require up to two full days and include steam cleaning of all surfaces and equipment, requiring that all food and supplies, mobile cabinets, and shelving units be removed. To ensure that restaurant and catering operations continued uninterrupted during this process, the museum could borrow one or more mobile catering kitchens from its current commercial caterer and rent trailers to store relocated nonperishable food, supplies, and light and mobile equipment.

All told, it typically takes about one day to clean out most kitchen and storage areas, two days for a professional facility cleaning, and another day to restock and return to normal operations. Though this is a significant time commitment and not an inconsiderable inconvenience for all involved, it is essential that all cultural institutions' foodservices do everything possible to operate in a consistently sanitary manner that promotes food safety.

Because administrators at this art museum needed to decide between retaining an enhanced self-operated restaurant and catering program or turning operations over to an outside contractor, two sets of relevant recommendations were provided. If remaining self-operated was preferred, the current program should be given an in-depth evaluation and assessment to develop a business plan to optimize current food and beverage services. This would include reviewing the restaurant's current operating hours, foodservice's organizational structure, and current management personnel's ability to attain museum-determined objectives. In addition, written policies and procedures for purchasing, employee relations, menu development, recipes, sanitation practices, catering services, and event booking would all have to be developed.

The key choices if a switch to contractor operation of foodservices was preferred were whether to initiate a search-and-selection process with the greatest possible discretion or to proceed publicly to expedite the transition. When privacy is most desired, contractors can be quietly asked to come in to review operations, provide an assessment and indication of interest, and outline the sort of financial terms and conditions under which they would be prepared to take over service management. There is always a risk that current staff will catch wind of a potential change, but it is commonly understood that organizations will periodically hold exploratory talks with outside service providers and that such discussions do not always result in changes of management. If the public search track is followed, the probable best-case time frame is two to three months, though three to five months is more common for this process. Since foodservice managers and chefs will hear of administrators' pursuit of a contract foodservice provider, financial incentives tied to minimum operating standards could be offered to ensure that these personnel remain until the new operator is selected and in place.

To help guide restaurant and catering growth over the short term, we advised the museum to begin comprehensive market research, including intercept surveys and telephone interviews with recent visitors, prospective visitors, and patrons of the restaurant and catering services. Research findings could serve as the basis for a new business plan (if foodservice remained self-operated) or to provide a selected contractor with information that would help the new operator serve all customer groups better.

Finally, based on the relatively high revenue foodservice was already then generating and this museum's willingness to learn to serve its markets better and further develop its hospitality programs, it seemed likely that a contractor would be willing to manage restaurant and catering operations on a profit-and-loss basis and pay the institution a commission on sales ranging from 8 to 10 percent or more, depending upon negotiated contractual and financial terms and conditions.

SUCCESSFULLY INTRODUCING A NEW RESTAURANT

At a museum of natural history with an annual visitor volume of approximately 250,000 (including festivals and special events), administrators' main concern was the successful implementation of a new on-site restaurant to be located in a planned extension of an existing building. The expectations shared by these administrators included high-quality foods and beverages, an innovative menu, and consistent quality control. A related objective was to ensure that the planned restaurant's concept, design, and menu approach all became integral parts of the museum experience, including recipes and dishes that drew from Native American, Mexican, and regional American cultures. Sound fiscal performance, which would allow this museum to realize maximal financial gain from its new dining facility *and* give an operator (and caterers) opportunities for profits, was also specified. Lastly, it was determined that catering here should be operated on a nonexclusive basis, though the new restaurant's operator would be expected to provide day-to-day internal catering services and would be promoted by this museum as the caterer of choice to existing and potential special-events clients.

At the time of our evaluations, certain aspects of the planned restaurant had been fully determined while others were still to be decided. What was certain was that the new restaurant's location would be in a courtyard across from the museum's $3.5-million-a-year retail store. Its area would include some 2,200 square feet, be rectangular in layout, and adjoin an outdoor seating patio, which, along with the courtyard, would provide another 2,000 square feet of dining space. Questions yet to be answered included whether the new restaurant's kitchen would be a full production and preparation operation requiring no off-site support or whether it should be a more limited facility capable of providing just meal prep, plating, and storage support. While it had been agreed that this restaurant's equipment would be generic in design, to allow the widest possible selection of potential operators to run their programs without impediment, it was then determined that meal services would include complete hot and cold luncheon service and light snack and beverage service during mornings and afternoons. In addition, the restaurant would be the source of all day-to-day catering production, supported by catering pantries in several special-events areas.

OPERATOR INVOLVEMENT AND FACILITY FINANCIAL PLANNING

One possible concern that arose while assessments were being made was whether the museum would be able to secure a restaurant operator capable of running (and willing to run) the planned facility on a nonsubsidized basis and offer a financial return to the institution without requiring exclusive control of catering. We recommended that if no suitable, qualified operator could be found who was willing to work under those terms, the museum should offer the most likely operator exclusivity on catering, but only with an agreed-upon exception that would reserve to the institution the right to hire donor-selected outside caterers or those more capable than the restaurant operator of handling the three to five largest (and most lucrative) special events. This would ensure that the selected caterers would face no competition for the vast majority of events.

We also noted that the museum would be best served if it bought and stored (space permitting) its own inventories of tables, chairs, and tabletop items to support on-site catered functions and provide a revenue-generating rental opportunity for off-premise special events.

With these operating conditions established, recommendations were offered to facilitate the planned restaurant's build-out. A decision to require the operator to build out the facility or provide all or a significant portion of the capital required to do so would most likely reduce interest in this project, especially since catering was to be (at least initially) offered nonexclusively and this museum had a relatively low annual paid attendance. Therefore, the best course appeared to be for the museum itself to fund the build-out (capital resources permitting) and provide the facility's operator with a fully equipped (save for smallwares, cash registers, office equipment, and the like) turnkey operation. The size of the investment required in kitchen equipment would then depend on whether the institution preferred to commission a fully self-supporting production center or a more modest facility backed by off-site meal production and storage.

Given these parameters, it was estimated that the cost of equipping the new restaurant (excluding general contractor services) would be between $100 and $150 per square foot or $220,000 to $330,000 all told, not counting the cost of the dining patio and its furnishings. One way these expenses could be reduced was through the purchase of used food-production and preparation equipment at auctions, provided the pieces were inspected, serviced, reconstructed, and in good operating condition (we do not recommend this approach, but it is an option. In addition, other equipment, such as ice machines and dishwashers, could easily be leased rather than purchased, and pieces such as cold beverage dispensers and coffeemakers are often provided at no cost by food and beverage suppliers. One final issue—which, like those listed above, could be decided during the final design process—was whether equipment should be purchased through the project's general contractor or directly from a dealer or distributor. While administrators at different cultural institutions may prefer one kitchen equipment purchasing method over the other for a variety of reasons, it is usually best to explore costs associated with both options before making a final decision.

Once museum administrators had decided which of the two types of kitchen was most suitable for their new restaurant project, we recommended that they engage a foodservice facility design consultant who would provide the final layout of the kitchen/serving portion of the operation and prepare a project budget. At the same time, an interior design firm would need to be retained to work with the foodservice facility design consultant in completing the interior and final design and budget. Concurrent with these steps, administrators should begin their RFP process, with a goal of selecting a restaurant operator *before* the design plans were completed in order to give the operator input on issues such as equipment location and placement of bussing stations and storage areas. It is worth stressing that in new construction (or renovation projects) where the operator is either chosen too late or otherwise not given a chance to offer input before design planning is finished, operational problems arising from suboptimal traffic flows or layout configurations are likely to occur. That's why coordinating suggestions from all project partners during the planning process can frequently result in decreased operating costs, thanks to a design that has been determined by all partners to be as efficient as possible.

Our concluding observation for this museum was that given its location, concept, and flexibility, the planned new restaurant had a high probability of successfully attaining identified objectives. What could not be determined, however, was whether success could be achieved if the operator did not have exclusive rights to catering. Therefore, it was stressed that the museum might well have to allow catering exclusivity for the sake of attaining its larger project goals.

CHOOSING APPROPRIATE MANAGEMENT FOR A NEW FACILITY

Another assessment of the best way to introduce a new restaurant was recently provided to a small zoo, whose administrators were concerned about the impact of this new facility on an existing snack bar operation. The evaluation process began as it usually does, by identifying administrators' objectives for their new operation, as well as the scope and condition of the overall foodservice program. The snack bar at this zoo, for example, was then producing average annual sales of about $180,000, returning 12 to 13 percent to the zoo (around $25,000 a year). Food and beverage carts and a gift shop were also in operation. Catering, however, was not a major income or sales component of this program and, though several local caterers were used for special events, no current sales data or projections for potential profitability had been developed. All told, with an attendance of about 260,000 annually, foodservices at this zoo were recording per capita revenue of about $0.70, which was well below the norm at most other U.S. zoos.

The planned restaurant was designed to operate amidst one of the zoo's continental displays and offer a self-service program. The facility's size and design had been modeled, according to the architect, on restaurants commonly found in other zoos, and was not themed around a single preparation style or concept. The design also called for a full kitchen, serving space, and a large, multipurpose indoor dining area, suitable for both customer seating and hosting special events at times when the zoo was not open to the public. It was estimated by the client that build-out for this restaurant would cost approximately $133,000 (including kitchen equipment) and about another $100,000 for final construction, interior design services, and furnishings and fixtures. The administrators' stated goals for this new foodservice facility were to offer customers a high level of convenience, fair and competitive menu prices, and a market-driven variety of item choices, as well as to earn new profits for their cultural institution.

Perhaps the most pressing issue still to be resolved at the time of our assessment was whether the new restaurant should be self-operated or outsourced. Based on our experience, the bottom-line income and potential profitability of a restaurant in medium- to large-size cultural institutions is about the same whether it is managed in-house or outsourced. When a contract foodservice provider is retained, however—and particularly when that company has had specialty experience operating in zoos or similar environments—our experience has shown that there is upside potential due to the fact that the contractor has greater purchasing power, market-proven operational systems, and promotional resources that can provide stronger opportunities to grow foodservice net revenue, which in turn can create more upside profit potential for both an institution and its contractor.

We have further found that at smaller cultural institutions, such as this zoo, with fewer than 300,000 annual visitors, outside food-service contractors usually can run a program more efficiently than self-operators, since they will have an experienced, highly qualified on-site manager, their staffing will likely be leaner, their customer service will tend to be quicker and more error-free, their more extensive menu variety will create larger average checks, and their promotional and merchandising programs will be more professional and market-driven. The upshot of these advantages is, typically, more revenue and profits for smaller client institutions.

It should also be emphasized that a self-operated restaurant at almost any cultural institution, including this zoo, can perform only up to the ability, qualifications, and experience of the manager in charge. In the case of a small zoo, it was unlikely that the institution could afford to employ a manager who would be as experienced or more experienced than a peer employed by a qualified contractor. What's more, the zoo's foodservice manager would have (as noted in Chapter 4) no upward career path within the institution. Thus if the zoo hired a young, ambitious manager for its planned restaurant, this person would probably soon no longer feel challenged by the assignment and would look to move on to larger and more exciting opportunities in the restaurant/hospitality industry. Hiring a restaurant manager closer to the end of his or her career, while offering the prospect of a more stable and long-term relationship, would subject the zoo to the risk that this manager would not be as motivated, innovative, or dedicated to creating excitement and enthusiasm among colleagues and visitors as would a contractor's employee.

In addition, as previously mentioned, most smaller cultural institutions, such as this zoo, cannot afford to provide the top salaries and competitive bonus-incentive-benefits packages that are commonly offered by commercial-sector restaurants and foodservice management companies. As a general rule, museums, zoos, aquariums, and other cultural institutions that self-operate restaurants and other foodservices usually have larger-than-average annual visitor counts (750,000+ annually), as well as per capita food expenditures two to three times greater than was the case at this zoo. This sort of operational scale better enables a self-operated foodservice to attract top-flight management, pay market-level salaries and benefits, and (to some extent) take advantage of economies of scale in areas such as purchasing.

Our evaluation at this zoo also led to the observation that catering and special events represented an untapped opportunity for additional new income, through both facility rental fees and profits from the sale of foods and beverages (regardless of whether the service was self-operated, contracted to an outside provider, or managed by an off-premises caterer). Because the zoo had yet to develop a full-scale special-events program, it enjoyed several expansion options. One would be to give all food and beverage catering on an exclusive basis to an on-site operator (whether as a zoo department or outsourced to a contract manager). A second choice was to reserve *only* alcoholic beverage catering exclusively for the on-site operator, thus assigning the licensing and insurance responsibilities to the institution. The third option was to open the special events to any caterer, including the on-site operator (or to create a list of approved caterers). However, this zoo chose to develop its catering and special-events program so it could anticipate revenue at least equal to 10 to 15 percent of food and beverage sales, as well as additional income from facility rental fees.

After sharing these assessments, based on the zoo's goals and priorities, we concluded that this zoo's best option was to outsource all its foodservice operations, though we added that the exclusive right to handle catering should be reserved until an outside operator with appropriate qualifications was retained. However, if the zoo did not receive one or more proposals from qualified outside foodservice providers who could reasonably demonstrate their ability to meet administrators' stated goals, self-operation should be maintained. To begin the necessary transition, the zoo was urged to begin its search for a qualified outside foodservice operator immediately. As has been noted, the more quickly an outside operator was hired, the more opportunity the foodservice provider would have to contribute to the final design of the planned new restaurant, maximizing the chance that it would operate at peak efficiency.

SOLICITING A NEW OPERATOR WITHOUT A RFP

During this particular consulting assignment, we did not recommend that a formal RFP process be initiated. This was because the zoo's potential maximum foodservice revenue of $400,000 to $700,000 per year would likely elicit bids from only a few conventional and experienced operators. Instead, we provided this zoo's administrators with a list of three qualified outside foodservice providers with experience managing restaurants and concessions in zoos. Because the zoo wanted to handle the RFP process internally, we recommended that administrators contact each company to determine its interest and to send whatever information was requested to the one(s) most eager to partner with their institution. Interested outside operators should be invited to tour the zoo's facilities, review its long-range plans, and submit a proposal. It was also advised that the zoo offer to fund the restaurant's build-out but inquire about the degree of financial support an operator might contribute. The zoo should further indicate that it would be most interested in a contract that offered a return percentage based on total gross food and beverage sales, but that it would consider other options. Finally, it was important to ascertain during these negotiations whether the operator(s) would require exclusive rights to on-premises catering as part of their arrangement with the zoo.

In addition, we recommended that zoo administrators concurrently contact local quick-service restaurant franchisees (such as

McDonald's, Burger King, Taco Bell, Subway, and Pizza Hut) to determine whether they were interested in installing a unit in the planned restaurant's space. The presence of two or three such branded concepts might well offer an attractive foodservice option for visitors. This effort could best be supported by additional contacts with large, experienced commercial catering firms in the local area to ascertain their degree of interest in partnering with the institution's management. All these efforts would let zoo administrators determine whether a qualified outside foodservice provider was available to manage their program on terms acceptable to the institution. If no such provider appeared, the zoo would then know for certain that continued self-operation of foodservices was its only remaining option. Should continued self-operation be required, we advised that this zoo bring on board (or select from its existing personnel pool) a foodservice manager as soon as possible. This would ensure that the person responsible for operating the planned restaurant would be involved in the project early enough to tweak and adjust the preliminary, generic design to fit selected menu and design concepts.

6

Developing and Marketing Special-Events Programs: Challenge-Solution Case Studies

Museums, zoos, aquariums, and other cultural institutions go through many permutations as they grow and change, responding to the ever-evolving interests, expectations, and tastes of their constituent audiences and sponsoring groups. Over the course of their productive lives, most cultural education and entertainment venues face a need to adjust their physical plants to respond to rising or falling guest volume, exhibits' scope, contents and intentions, and new display and information-sharing technologies, to cite just a few change factors.

Because of Americans' continually growing interest in informative and entertaining displays of cultural and historic artifacts, as well as their increased disposable income and leisure time, cultural institutions have been able to market their collections to increasingly broader population groups. These growing audiences have meant that many such organizations have recently renovated or expanded current buildings or embarked upon new construction projects.

While exhibit and open space (lobby, great hall, etc.) is always paramount in administrators' minds when facilities are redone or built, sufficient consideration is not always accorded to the business opportunities these spaces offer as sites for expanded or entirely new special-events programs. Incorporating enhanced special-events and catering programs into facility use planning and, especially, the design of new or remodeled facilities can offer substantial financial rewards to cultural institutions for many years to come. Well-integrated and properly promoted special events and catering can raise community awareness, spur support for new or redone exhibit areas and their contents, and serve effectively to market a cultural institution as a destination of choice for first-class hospitality services. Internal catered functions can be more easily accommodated and efficiently served when their requirements are incorporated into planning for new spaces. Just as important, properly managed special-events programs can provide an institution with funds that can be used for purposes ranging from paying for new construction to purchasing desired collections or offsetting decreases in government support. In addition, the more revenue a special-events and catering program can earn from food and beverage sales and facility rental fees, the more on-site restaurant programs can afford to be customer-focused, diminishing their need to be concerned with earning profits at the risk of alienating guests with high costs or inferior service or quality.

Special-events services are increasing in importance for many museums, zoos, and other cultural institutions, in part because public demand has been growing for meal and entertainment functions in the sort of highly aesthetic, historic, and/or educational environments traditionally found in museums and other cultural institutions. It is thus imperative for administrators to coordinate facility rental schedules and fee structures with the development and promotion of dining services (as well as housekeeping, security and other support services) when a new or expanded special-events program is in the offing.

While reading the following case studies, remember that commercial-sector special-events standards, presentation and service techniques, and marketing strategies may all be worthy of emulation by catering programs in cultural institutions. This awareness of the best commercial catering practices can be particularly useful for museums and other institutions that have earned reputations as the premier venues of their type in their communities.

Building Hospitality into a New Space

At one medium-sized art museum, resources included an impressive collection of buildings and open spaces in a campus environment that was about to be augmented by the opening of a new building. Administrators had already determined that the first priority for use of the new space would be to host internal catered functions, but a significant number and variety of external special events could be accommodated, as well. This finding was especially significant because previously this museum had not been readily available as a special-events venue. The cachet of the museum's collections, combined with the availability of the new facility, however, indicated that it would make the museum an attractive site for individuals and organizations looking to hold catered receptions and the like.

Timing was another important consideration when it came to planning the development and promotion of special events here. The new wing's spaces would be available for special events only after a very popular, high-traffic, temporary exhibit had finished its run but, because many corporate and civic organizations (as well as families) plan their special events far in advance, there was still time to solicit and schedule a new roster of events. However, we emphasized to the institution's administrators that establishing the museum as the community's catering destination of choice would take a considerable amount of time.

DEVELOPING SPECIAL-EVENTS MARKETING

To begin the development of an expanded special-events program, we recommended the establishment of a sales and marketing initiative that would present the museum's newly available spaces to a variety of potential customers. In this institution's locale, these included meeting planners, public and private event coordinators, and corporate executives and sales managers, among others. We also strongly urged that the expanded special-events program be packaged and presented at the high end of the local market, both to ensure that all events reflected the excellent hospitality and aesthetic standards already established by this museum and further to reinforce its reputation among catering customers, both internal and external. If these special-events standards were consistently applied, it was likely that the administration would soon come to see the expanded program as an integral part of the museum's total services package, one especially valued for its ability to earn new gross revenue and profits for the institution. The final preliminary recommendation was to benchmark potential events' pricing, space usage requirements, costs, and fees against those of the best local commercial catering firms, and to set prices that both demonstrated the advantages offered to museum members and encouraged nonmembers to rent the new spaces.

With preliminary special-events program outlines determined, the museum next needed to begin the business planning that would be required to guide the growth of the expanded program. The first step was to prepare a preliminary estimate of special-events revenue. This was based on estimates of the market value of the museum's available spaces (compared with the cost of existing commercial catering venues) number of guests served, and competitive meal/reception pricing multiplied by the number of events likely to be scheduled during any calendar period. Next, we advised that this estimate should be supported by an identification of the resources necessary to generate projected revenues. Specifically, administrators needed to determine the extent of the marketing programs, staff, fulfillment personnel, and additional organizational structure required to develop and manage the anticipated calendar of special events, and how much these resources would cost annually, to ensure that sufficient profit would be generated to justify the institution's investment.

It was also important to include all potential revenue sources and assumptions about income generation in any preliminary estimate. In the case of this museum, it was agreed that some new spaces would be open to special events attended by both members and nonmembers, while some areas would be reserved strictly for members' dinners or parties, and others would be restricted to nonmember use. To maximize the tie-in between this cultural institution and its special-events program, wherever and whenever possible the new wing's spaces would be decorated with museum objects or themes related to the artwork. Marketing of new special-events opportunities would commence during the summer months to raise community awareness of museum hospitality resources before the next holiday season, as well as to gain initial new business that fall. However, this would require the museum to hire all necessary new event marketing and administrative staff by late spring.

While it was acknowledged that the new wing's spaces would host a busy calendar of internal, institution-sponsored special events, museum administrators also expected that the new marketing staff would assemble a calendar of the most desirable outside-funded events to add to the existing roster. In addition, before new events were solicited or booked, the special-events staff would be expected to review membership costs and discuss with membership department personnel the best ways to market the value of the institution's corporate membership program. This would assist with the establishment of facility rental fees, given the requirements that these charges be locally competitive and offer different charges to members and nonmembers, yet reflect the museum's position as a premier special-events location. To ensure maximal special-events revenue from the new wing, we suggested further that each space be given its own fee structure, adjusted according to room capacity, attractiveness, and likelihood of rental, in order to encourage maximal facility utilization. We also pointed out the benefit of welcoming special-events business from selected local nonprofit organizations, since although such groups would be offered catering services and facility rentals at an appropriate discount, the goodwill and positive publicity their events would generate would likely more than make up for the reduced revenue.

It is very useful to keep in mind that all preliminary revenue estimates and assumptions will need to be adjusted to reflect changes in the local market, competitors' offerings, the state of the regional or national economy, and the quality of service, as well as other relevant factors. This is why such estimates should be considered as best-available initial inputs intended to guide the development of a new or expanded special-events program and should be subject to revision as actual bookings are recorded and compared against plan projections.

BUILDING FINANCIAL MODELS

Next, we considered assumptions pertaining to the volume, location, and revenue to be derived from museum-sponsored and corporate-member special events. The two existing spaces at this museum were currently hosting nearly 40 members-only special events a year—less than two-thirds of their capacity. Growth projections estimated a 100 percent increase in scheduled events over the next two years (capacity would increase as the new wing's spaces became available), with nonprofit groups' bookings constituting some 10 percent of this higher total, as the growth in corporate-sponsored events would likely influence nonprofits to patronize the museum's catering services. With this projection in hand, it was decided to leave event and facility rental fees unchanged for the existing spaces, both to provide an incentive to local companies to become corporate members and to allow future sales efforts to focus on increasing space usage rather than justifying increased costs. Calculations were next performed to determine total annual special-events revenue before expenses, and included the 35 percent discount to be offered to nonprofit groups.

Similar revenue estimates were subsequently worked up for all this museum's available special-events spaces (including those available to nonmembers as well as institution staff). Usage rate increases were projected, allowing administrators here to consider whether the spaces that were potentially the most appealing should charge lower fees in order to attract the widest possible group of users from the local community, and whether it would be more advantageous to try to book a number of small groups into large auditorium spaces.

Because all calculations were estimates based on current usage rates, as well as the potential of the museum's current and new spaces as special-events venues, and our experience with the results commonly attained by new or expanded special-events marketing programs, three revenue models were offered to the museum's administrators. These included a set of "most likely" midpath revenue projections, a high-path set (midpath plus 10 percent), and a low-path set (midpath minus 20 percent). We advised that for the next two years, administrators should base their revenue assumptions on low-path figures, while allowing for a modest 10 percent increase above low-path estimates as the newness of the museum's catering spaces ceased to be a compelling sales inducement (which likely would cause usage to level off).

Other factors that needed to be determined at this time included whether to establish security and other staffing expenses as a component of general space rental fees or to break them out on a fluctuating cost-per-space basis to cover this incremental expense as closely as possible. We also advised that a cost accounting system be developed that would allow actual costs to be recorded for each space according to the size and nature of each special event, support net revenue projections, and help event-management staff create financial reports. To keep operating cost estimates consistent with revenue projections, it was recommended that anticipated personnel, marketing, and related expenses be budgeted against low-path figures to ensure that sufficient return would be preserved for the museum and that staff and other resources would be enhanced only insofar as special-events volume and revenue warranted.

DETERMINING AN EXPANDED STRUCTURE

In addition to existing fulfillment staff, charged with ensuring that events were operated according to guests' expectations and museum standards, new personnel responsible for initiating contacts with potential clients and selling new and established event resources would have to be hired. This museum was fortunate that its fulfillment staff had already proved themselves capable of coordinating all the details required to stage events as intended; thus, we recommended that the special-events department add a junior-level person to the operating team, to provide administrative backup and allow current staff to manage more events without feeling overstretched. (We noted that as special-events volume increased during the next two years, more fulfillment personnel would likely be required; provisions were consequently made in the department's table of organization.)

Perhaps the most significant personnel finding was that there was no one on the events department staff capable of leading an expanded organization and assuming responsibility both for marketing and sales and for increased administrative coordination. Therefore, we urged that an experienced senior department head be hired. In addition, we identified the need to add a full-time sales manager, one capable of assisting the department head's selling efforts.

To ascertain whether such a staffing structure would allow museum special-events administrators to return sufficient profits to their institution from the projected revenue streams, all existing and estimated staff-related expenses were calculated. Planners were careful to include all department employees' base salaries, fringe benefits, and potential bonuses in their expense charts, noting that if high-path sales figures were in fact attained, potential bonuses could increase by as much as 15 percent above projected midpath performance levels. Additional cost lines were dedicated to office expenses, travel and entertainment in support of new sales, and such marketing materials as an events-promotion booklet, a revised mailer, a sales presentation kit, and advertisements in news media.

One important element influencing the planned expansion of this museum's special-events program was its future relationship with its on-site restaurant operator and caterer. One suggestion offered was that this operator might provide some or all of the new staff required to carry out the additional duties and service responsibilities envisioned for the enlarged department. It should be noted that a number of prominent U.S. cultural institutions, including the Museum of Science and Industry and the Shedd Aquarium, receive much of their special-events staff support from their outsourced on-site exclusive restaurant and catering providers.

SETTING A MARKETING STRATEGY

The final, though hardly least consequential, element to be determined in support of this museum's special-events expansion was its overall marketing plan, including marketing strategy and tactics for its implementation. While final details were left to be worked out with the still-to-be-hired department head, the proposed marketing strategy consisted of four objectives: to make prospective special-events customers aware of what this museum had to offer, to contact them directly to gauge interest in booking events, to convince them to visit the special-events spaces to see how well they met their expectations, and to sign prospects up as clients.

To execute this strategy, the first step was to identify and prioritize prospects. Prominent among these were current museum members, local corporations, professional associations, nonprofit organizations, convention attendees, tour groups, families planning social functions, community organizations, out-of-area companies that held meetings in this institution's locale, and both local and out-of-area art aficionados.

The second recommended step was to prepare new and revised sales materials. The theme of these materials would be that this museum had now become the premier community venue for special events of all types. It was important that sales pieces not only capitalize on the institution's well-established art reputation, but also emphasize its newly enlarged capacity to host catered parties, meals, or receptions in an expanded variety of appealing and unique spaces. Materials such as brochures and flyers could best be supported by advertisements placed with selected news media and a direct-mail campaign targeted to prospects (including preaddressed reply or "call-this-number" cards).

We pointed out that enticing as many prospects as possible to visit the museum and tour its special-events facilities was a necessity for a successfully expanded program. Such tours would give potential clients a chance to experience the facility's decor and ambiance, taste prospective menu items, and become involved in developing their own events. To encourage visits by prospects, we suggested that the museum initiate a series of special lunches, receptions, and dinners exclusively for potential major special-events clients.

The fourth step, signing up new special-events clients, would need to be managed according to a formal closing procedure. Given this institution's public-service mission (shared by most cultural institutions), it was understood that closing inducements would have to be discreet but still apply the principles of good salesmanship. Suggestions included offering preferred space-reservation status and early-sign-up rewards such as pricing incentives or a choice of free "bonus" food-and-beverage items.

The last phase in our recommendations for developing an expanded special-events program here was the drafting of a projected profit-and-loss summary. This document laid out the reasons why this museum's program was likely to grow, steps necessary to achieve this growth, estimated revenue and costs for the first full year of expanded operations, gross profits before event costs, probable cost of the predicted number of events, and, last but not least, the projected net profits accruing to the institution.

MAXIMIZING MARKETING REVENUE IN A NEW VENUE

Another cultural institution, in this case a science museum, was recently facing a similar need to market and develop an enhanced special-events program as administrators awaited the opening of a dramatic new building. Our intent in this instance was not only to help identify the goals and plan the steps necessary for the successful expansion of special events here, but also to recommend ways to maximize incremental revenue realized from facility rental fees and a previously established percentage of food-and-beverage sales.

We began by identifying museum personnel's current expectations and assumptions. First, it seemed most logical to expect that the publicity and "buzz" surrounding the new museum building's opening would cause its special-events spaces to operate at capacity for the subsequent two to three years. One purpose of our plan, therefore, was to maintain popular interest in holding special events at this museum from that point forward. The expanded program was also expected to ensure that all affairs were staged in a manner that was consistent with the quality of the state-of-the-art new building and enhanced the museum's reputation as an unsurpassed provider of special events.

Because we were working with an extraordinary cultural institution, it was important that the general marketing strategy include an extraordinary sales approach. This decision was based on the capabilities of the new building, which featured an innovative exhibit display scheme dedicated to promulgating information on the sciences, mathematics, and engineering, as well as interactive installations including a fabulous variety of special event rooms and spaces specifically designed for this purpose. This suggested that the museum's special-events program could be productively marketed to a range of potential client groups that included companies, organizations, and associations involved with engineering, mathematics, and science; groups already patronizing this museum on a regular basis; museum donors; local civic and business organizations with a record of supporting cultural institutions; family groups seeking science-based "infotainment" for adults and children; and regular attendees at community education events. The recommended selling strategy to turn these potential special-events users into customers suggested focusing on groups that were large enough or that visited the museum frequently enough to warrant the expense associated with new marketing efforts. Targeted groups would be approached via direct sales calls supported by direct-mail campaigns in order to acquaint contact persons with the museum's enhanced special-events resources and establish ongoing relationships. If such sales calls and contacts were conducted professionally, repeat special-events business would likely be ensured and successive events scheduled all but automatically.

ESTIMATING REVENUE FROM EXPANDED SPECIAL EVENTS

To determine potential revenue for the expanded special-events program's first full year of operation, it was essential to identify each rentable venue within the museum, create a preliminary estimate of venue fees and the number of times each could be rented annually, and estimate all food-and-beverage-related income, based on the established percentage being paid to this institution by its caterer. It was then necessary to create an appropriate range of financial models to lay out potential revenue scenarios for both full-fee, full-occupancy business volume and events for which reduced fees would be charged and/or which failed to sell out rentable spaces. In all scenarios, it was understood that projected revenue was being estimated before incremental, nonbilled expenses such as increased maintenance, utilities, and staffing requirements were figured in.

Preliminary calculations indicated two reasons why administrators at this museum had reason to be optimistic about the growth potential of their special-events program. The first was the fact that the soon-to-be-opened building would be the newest of its type in its geographic region and would therefore likely immediately become *the* venue for local special events. Another advantage was that the new building had an unusually large number of potential special-events spaces available for a facility of its size, warranting an aggressive marketing program and offering the opportunity for a large net increase in income from hospitality services. Projections suggested that once the new building opened, annual special-events gross revenue accruing to this museum would range from a high of about $900,000 to a low of around $375,000. Therefore, it seemed reasonable to project first-year accrued revenue of approximately $400,000, providing a benchmark for the setting of operating budgets and a marketing plan. As is often the case when entrepreneurial enterprises are begun, following the more conservative set of income expectations seemed most advisable, especially since budgets and plans could be adjusted upward if and when program performance seemed to justify revisions.

In order to validate the various financial estimations calculated earlier, we created charts showing each potential venue's capacity, fees, average food-and-beverage charges, bookings per year, and annual projected revenue for both full-fee/full-capacity and discount/less-than-full-capacity scenarios. These charts showed in detail how different venues would perform depending upon their booking frequency and fee rate, providing a useful comparative tool to assess actual future performance versus estimates.

The next step in developing special-events operations for this museum's new facility was to determine the appropriate level of food and service quality. One complication was that the large number of available special-events spaces made it necessary to prepare to host different types of events, requiring a variety of menus, prices, and service levels. To ensure that all events would be appropriately handled, museum administrators were advised both to contract an exclusive arrangement with a caterer and to put into place internal staff qualified to manage fulfillment responsibilities including scheduling and coordinating event management with the selected caterer. From a marketing perspective, it was also vital to position the new special-events program to indicate that the museum was a premier hospitality venue, though not an unaffordable or "stuffy" one—in short, a destination worthy of guests' repeat business. Above-average food quality and menu prices (by local standards) were the most logical choices to instill this image, with casual-dining menu items to be specified at the top of the available quality range and fine-dining menus to be equal to those of high-end community restaurants.

The special-events expansion team's next task was deciding on the composition of the internal special-events staff. Of immediate concern was establishing the duties of the still-to-be-hired sales manager and sales staff. Among the manager's projected responsibilities were handling all bookings, generating repeat engagements, and ensuring that all institutional commitments to special-events customers were fulfilled. While this manager would have to be, above all, a sales professional, his or her job would also include serving as primary liaison to the outside caterer, coordinating food and service presentations to accord with guest expectations. It seemed evident that once the expanded special-events program got up to speed, the sales manager would need assistance from one or more sales representatives. The representative(s) would split work time between phone sales efforts and completing administrative work, such as ensuring that accurate and timely event records were consistently maintained.

It was also necessary to determine appropriate reporting procedures. Because almost all institutional special-events programs (including the one at this museum) are not part of core activities and manage their own revenue streams, operational difficulties and conflicts of perceived interests can arise around their function. Above all, it is essential that special-events programs properly balance the service needs of their guests, the needs of the parent organization, the needs of foodservice providers, and their own bottom lines. That's why it is strongly recommended that the new sales manager for special events be allowed to report directly to a senior administrator at the museum, preferably an executive capable of sorting out conflicting departmental interests and authoritatively deciding upon preferred courses of action.

SETTING POLICIES TO SUPPORT NEW STAFF

Following the establishment of these positions and procedures, it was important to create policies for marketing, coordinating, and managing forthcoming special events. These needed to be supported by the development of an administrative structure that would track client requests and satisfaction with event performance and maintain records and statistics. In addition, the performance of the new sales manager would undoubtedly be better if that individual was guided by formally drafted procedures for managing the relationships between the outside caterer, special-events clients, and the museum's own staff members.

As keeper of the master events calendar, a sales manager must be prepared to receive and provide input to all parties involved, including colleagues, donors, external customers brought in by the sales department, outside catering personnel, and managers and staff from institutional support departments. To carry out event coordination and marketing effectively, a sales manager should be aided by representatives or assistants with both good people skills and administrative ability. Therefore, individuals considered for such positions should be able to schedule events, update the master calendar, and help compile policy manuals, as well as successfully solicit new sales by phone, including placing cold calls, calls to donors, and calls to current clients likely to offer repeat business.

PREPARING A SALES PLAN

With these criteria spelled out, we looked next at the development of a sales plan. Prior to the new facility's opening, the museum's exclusive caterer had to be hired and menus tested, costed, and agreed to. All venue fees would need to be established, and it was essential that a museum representative (preferably the special-events sales manager) be in place to respond to event inquiries and initiate the booking schedule. While these duties could be handled by another of the institution's managers on an interim basis if cost savings were essential or if the hiring process ran into difficulties, our experience showed that it would be far more effective to bring the special-events sales manager on board well ahead of program expansions or debuts. That way, the manager would be prepared to handle all bookings, oversee the preparation of marketing materials, help to plan grand-opening activities, and establish a strong enough sales process that event volume would decline minimally or not at all after the novelty of the new facility began to wear off.

This last responsibility, maintaining strong sales in the months and years after a new facility's opening, often provides a true measure of a special-events manager's abilities. Because so many museum-worthy events are contracted many months in advance, no cultural institution's catering department can expect to thrive on new business generated immediately after a facility opening alone. To maintain a full calendar of booked rooms into the future, a special-events sales manager must develop a large and varied prospect list, as well as a selling strategy that perpetuates relationships with current clients via phone and personal sales calls and pursues lapsed and new relationships. As noted previously, all calls, personal contacts, and facility visits should be supported by a modest amount of advertising in media directed to meeting planners and other regular event customers, who should also be targeted as recipients of sales brochures and direct-mail promotions.

In figuring this museum's first-year expense schedule for its expanded special–events program, we estimated that the salaries, benefits, and (where applicable) bonuses for the new sales manager and one representative/assistant, combined with the cost of developing sales materials, mailing, advertisements and ad placements, and travel and entertainment required to promote the new facility as a premier special-events venue, would still be no more than 20 percent of the low-path estimate for total annual revenue. This would leave the museum sufficient new revenue to add more special-events sales representatives/assistants, as well as a marketing consultant, as event volume increased.

INITIATING SPECIAL-EVENTS MARKETING

Unlike the cultural institution cited in the prior case study, a metropolitan zoo had never been marketed and, thus, was not locally known as a desirable venue for special events. In addition, no formal procedures for marketing, coordinating, and staging private functions had ever been drafted, and this zoo had no administrative structure capable of tracking clients, maintaining records, or soliciting new event business. As a result, the challenges faced in this instance were to determine the best way to raise awareness systematically of the zoo as an appealing environment for special events, as well as present a sustainable and profitable events program.

To gauge this institution's resources and liabilities as a special-events site, a comparative analysis was undertaken. On the plus side the zoo's outdoor exhibition areas had a large capacity and had virtually no equivalent competitors in the local area. The zoo's proximity to other public entertainment sites and its central metropolitan location also afforded it good traffic flow. The institution had a well-known and unrivaled identity in its community, staff were used to working together in effective teams, and its picturesque and interesting environs gave it strong "location value" as a setting for special events.

Obstacles in the way of establishing this zoo as a first-choice events site appeared to include its lack of a foodservice reputation, venue use restrictions (including times and types of events and sound-level limitations), need for renovation or repair, minimal suitable interior spaces, and perceived problems with the incumbent caterer.

Supporting the zoo's desire to develop its special-events business, despite the above shortcomings, was the relatively large number of identifiable target clients. These fell into two categories: single-event customers and potential repeat customers. The most likely single-time clients were corporations in white-collar industries and manufacturing, nonprofit organizations, public sector employee associations, and education-affiliated groups. The most likely sources of repeat business were event and meeting planners, private party planners, tourist groups, and the zoo's own trustees, members, and guests.

Once zoo administrators were ready to address the creation of a special-events marketing strategy, it was agreed that targeted potential clients would be primarily contacted first through direct sales calls, and only later by direct-mailed marketing materials. It was felt that if the zoo's special-events representatives were able to establish personal relationships with new prospects, the program would best be able to meet its goal of offering extraordinary events in an extraordinary venue. Specific objectives for marketing special events included establishing the zoo's reputation as a private function site of choice; ensuring that the program would meet or exceed comparable standards in all phases of operations, especially guest service and marketing materials; making continuous improvement a fundamental tenet of the program, with the expectation that it would result in a higher client retention rate; increasing revenue substantially and quickly; and increasing targeted guest groups' awareness of the institution, its exhibits, and educational programs.

CREATING ACTION STEPS

The setting of objectives was followed by a listing of action steps required for their achievement. Two steps involving the sales process included attaining cold-call goals and meeting directly with end users (potential guests) to identify their special-events preferences. Other steps called for bringing prospects to the zoo's grounds for site tours as a means of visualizing their events; initiating a limited direct-mail program promoting the zoo as an excellent site for corporate picnics, holiday parties, and other sorts of events; establishing client/prospect databases; researching competitors' products and prices; and becoming better informed about the hospitality services industry. The final recommended step was to categorize all accumulated data so that prospect lists could be broken down into the categories of "primary" (most likely to book events), "secondary" (less likely to book events), and "tertiary" (all others).

To make sales calls and mailings as effective as possible, it was important to develop an attractive, multipart sales kit. We recommended that materials include descriptions of all available event sites, photographs of recent on-site events and "animal encounters," and promotional text emphasizing the zoo's unique resources and available "experiences." Inexpensive, zoo-themed gift items could also be created and mailed to accompany initial contact letters or notes of appreciation after site tours or events. We further advised that selling efforts should be initially aimed at the perceived "primary" and "secondary" targeted prospect groups, with the goal of booking events of 100 to 500 guests (leaving larger parties till after the program was better established). Early business could also be encouraged if zoo administrators prepared to exhibit at and attend local meeting planners' conferences, promote special-events capabilities to the community's visitors and convention bureau, and establish contacts with other organizations that could contribute to increasing the number of events held at the institution.

One usually effective way to debut a new or expanded special-events program is to hold a cocktail reception for such "event influencers" as meeting and party planners, media and entertainment executives, private caterers, and others able to communicate with large numbers of potential clients. Such receptions typically include facility tours, and their expenses are frequently shared with the outside caterer (especially if this company has an exclusive arrangement).

To make sure program development and promotional events took place as planned, we advised that the special-events sales director spend 60 percent of work time on selling tasks, including calls on prospects, research, site inspections, networking, and overseeing advertising and promotions. Another 30 percent of this director's efforts could then be dedicated to coordinating quality control systems and relations with the institution's various governing groups, as well as attending events, leaving 10 percent available for administrative tasks required by day-to-day operations. When commencing or expanding a cultural institution's special-events program, it is essential that sales personnel do at least as much selling as their planned schedule requires, so that projected revenue targets will be reached. Establishing these targets from scratch—as was the case at this zoo, since no event income records had previously been kept—is always something of an iffy business. We began by suggesting that during the zoo's first year of fully marketed special events, while the program was getting up to speed, revenue records should be kept on an annual basis, switching to a fiscal year accounting system at a later date. It was further estimated that this zoo would host some 75 events during the program's first year, average 200 guests per event, and levy an average food-and-beverage charge of $35 per person. Average site fees were estimated at $2,000 per event. This estimated total annual revenue and event total were projected to rise by between 10 percent and 12 percent during the second year of expanded operations, potentially earning this zoo a little less than $250,000 in new net revenue.

In this instance, expenses associated with reaching program objectives and revenue targets, such as the sales director's salary (and potential bonus), advertising, and development of a sales kit, projected out to less than 60 percent of estimated revenue, implying welcome new income (and repute) for the zoo.

PREPARING (AND USING) AN IMPLEMENTATION CALENDAR

The final major component of this special-events expansion and upgrade plan was to create a nine-month implementation calendar. Key goals for month one included agreement on all standard menus to be provided by the outside caterer, initiation of development of the sales kit, approval of standard contracts and agreements, development of a filing and records system, and the beginning of database compilation. By the end of the second month, it was expected that cold calls and direct mailings would be promoting the zoo as a site for corporate picnics, that plans for the event planners' reception would be well under way, and that sales kit materials would be completed and approved. During month three, event planners would receive invitations to the reception to be held in month four, and corporate picnic bookings would continue to be pursued. By month eight, direct mailings and cold calls would be used to promote the zoo as a venue for upcoming holiday parties, with the first holiday event bookings expected to be recorded during month nine.

The achievement of these program enhancements *and* increased event activity would most likely result in an improvement in this zoo's image and the growth of its donor base. One key remaining variable, however, was how to help improve the image of the incumbent on-site caterer (in this case, a contract management company) in the minds of special-events bookers and potential guests. We suggested that the on-site caterer develop and market a new business identity complete with separate logo, phone lines, and business cards.

Taken collectively, the steps of this special-events expansion plan gave this zoo's administrators a chance to generate significant new revenue, increase institutional visibility, and raise customer satisfaction to new levels by stressing quality. Satisfied event customers, in turn, would help to guarantee growing repeat business, as they would be likely to return and to recommend the hospitality service to colleagues and friends.

TAPPING INTO NEW MARKETING OPPORTUNITIES

At a small though prestigious art museum, administrators were facing the challenge of determining which special-events spaces would be most in demand and what sort of revenue and operating conditions should be anticipated when a new building opened on their campus several years in the future. These administrators were also strongly concerned with how new special events could best be marketed, the sorts of yet-untapped market opportunities the program could capitalize on, and how enhanced hospitality services could improve the cultural institution's (somewhat elitist) reputation in its local community.

To address these issues, we recommended that the museum include as many special-events spaces in its planned new building as possible, thus maximizing potential benefits to members, as well as guest volume and event revenue. Regardless of how extensive the expansion of the special-events program was to be, however, we strongly advised that the new building's events operations be developed to mesh well with the museum's infrastructure and express both its collections' themes and the institution's reputation for first-quality cultural education.

Because the timeline for the new building's introduction was still sufficiently long to allow specific special-events areas to be planned, we drew up a list of multifunctional rooms that, in our experience, were most capable of generating maximum revenue and attracting bookings from the widest possible array of special-events guests. This list included an auditorium with about 130 seats that could be rented to local businesses, provide a flexible venue for museum art department activities, and be converted to a banquet space as required. Ideally, this auditorium would be situated adjacent to a multipurpose banquet/reception space capable of hosting up to 300 guests for sit-down meals and as many as 500 for cocktails. One immediate benefit this new area would offer the museum would be the ability to reserve an existing sculpture garden just for member-sponsored events, thereby promoting membership growth and new donations. On its own, the banquet/reception space would provide a prestigious location available to any local company or organization able to pay the facility rental fee and meet requirements for minimum number of guests and per-person meal charges. It was recommended that, if at all possible, this space be separated from the suggested auditorium by a removable wall, allowing the combined areas to accommodate even the largest anticipated events. In addition, since a courtyard was planned to adjoin the new building, it was important to provide access from the banquet/reception space to create an indoor-outdoor option for event planners and guests. Given the attractiveness and flexibility of the recommended banquet/reception space, museum administrators would also need to decide if weddings and other private social functions would be solicited or even permitted.

The third and final special-events space recommended for this museum's new building was a small, upscale dining room dedicated exclusively to private dinners for senior corporate managers, institutional trustees and senior staff, and VIP visitors or guests of the museum. We advised that this dining room should have its own manager and kitchen staff capable of consistently providing meals and services of the highest standard, offer a separate gourmet menu, and provide seating for no more than 30 guests. Again, it was suggested that the dining room be expandable, with a removable wall, to allow seating capacity to be doubled should the need arise, but that the space should be marketed in its 30-seat configuration to promote its special and intimate dining ambiance.

With these new spaces and their purposes providing special-events program development guidance, it was appropriate to establish incremental revenue projections. Facility rental fees, in a suggested range of $500 to $2,500 per event, were determined after a review of comparable event space charges in the local area. As a premier venue, this museum had the responsibility of ensuring that special-events menus presented the highest-quality foods and beverages, but it also had a commensurate opportunity to price event meals accordingly. This meant that administrators here could expect commissions from outside caterers for external events of 10 to 15 percent of the caterers' revenue and, if alcoholic beverage service was licensed to and operated by the museum, the institution would see additional profit on this portion of revenue. Combined facilities fees and food-and-beverage income would, therefore, likely compute to between 35 and 40 percent of total user expenditures on special events, which, in the case of this institution, came out to roughly $80,000, derived from an estimated 145 events during the first full year of expanded operations.

MARKETING LOCATION VALUE

To achieve these returns and volume objectives, it was necessary to base special-events marketing plans upon the new building's apparent advantages as a hospitality destination. Because the museum's new facility featured an intriguing design and a well-known institutional affiliation, we expected that it would appeal to a broad range of regional business, civic, and educational groups. Because many of these organizations could reasonably be expected to be repeat special-events users, dedicated sales and service follow-up mechanisms would have to be put into place. Development of an enhanced special-events sales and service structure would further allow this museum to pursue business and build relationships with groups coming to its home city for one-time or annual events (such as business meetings or association conventions).

Pursuit of a broad range of local and national special-events clients can help a cultural institution create a market situation in which demand outstrips supply. That is, by making sufficient but limited spaces available for special-events bookings, even high-end rental fees and menu prices are most likely to be seen as justified by customers. That is why administrators' goal should be not to see how much special-events space they can offer, but to market *just enough* space to keep a certain amount of demand unmet.

Ensuring that the desired demand in fact materialized would require a two-phase special-events marketing program. We recommended that press coverage and direct mailings be employed before the new building and its rentable spaces opened to heighten public awareness of the museum's enhanced hospitality resources. Then, once the new facility was available to visitors, direct sales calls and a modicum of advertising should be added to widen the pool of potential prospects.

While marketing plans and materials were being developed, it was also important to determine effective communication schedules and clear lines of responsibility relative to the management of the enhanced special-events program. To that end, regular special-events management meetings, chaired by a senior staff person (preferably the director of operations), should be held several times a month (or more often, if problems remained unresolved between meetings). In addition, a special-events master calendar would need to be developed; this would prove most useful at maintaining effective interdepartmental communications if its database software would allow real-time access by members of both the special-events and membership departments. Given the interelationships of multiple departments that inevitably occurs whenever a special-events department is initiated or expanded, we advised that the special-events department's personnel ultimately report to the institution's director of finance and operations. Placing an administrator with significant responsibility for (and experience with) business activities in charge of special events was most likely to result in the program being recognized as a revenue-producing, profit-making enterprise, and being run as such.

DEPARTMENT DIRECTOR'S ROLE DURING EXPANSION

The key individual responsible for the day-to-day development and management of special events would, of course, be the department's director. In the case of this museum, we recommended that during the period directly preceding and following the new building's opening, the director should assume responsibility for selling events, overseeing fulfillment obligations carried out by the foodservice provider and internal facility-support personnel, and coordinating event marketing, scheduling, and execution with the institution's membership department. The special-events director would also have to evaluate and recommend all menu pricing and á la carte food-and-beverage charges, ensure via publicity and direct mailings that targeted user groups were aware that new spaces were available for special events, accept inquiries and orders, and manage reservations and any other functions necessary to ensure the provision of services.

We expected that by the time the museum's expanded special-events program had been operational for a full year, the department's director would be able to divide daily duties, spending half of work time conserving current business and reselling existing clients to create repeat business, and the other half on marketing efforts intended to provide a full events calendar in the future. It seemed likely that at some point (determined by the growth of special-events volume) an administrative assistant would have to be hired to support the department director. This individual would be expected to take over responsibility for the master calendar and handle billing, pricing, and internal fulfillment services for all special events. A good rule of thumb is to budget the compensation

package for a director of special events, including salary, fringe benefits, and commissions (payable only after the completion of an
event and receipt of the client's payment in full), at about two and
a half times the amount to be paid to the administrative assistant.
(Actual dollar amounts will depend on local wage scales and labor
markets.)

The creation of a sales kit and brochures, which should include
regularly updated descriptions of all venues, menus, prices, and
other "must-have" information, can be expected to cost about
$25,000, though this is largely a one-time expense. Additional sales
materials, travel and entertainment expenditures, direct-mail pieces,
and mailing charges can all be expected to total somewhat less than
$10,000 annually, depending on the size of a cultural institution's
target markets and number of projected user groups.

CHAPTER 7

Performing Special-Events Department Assessments: Challenge-Solution Case Study

When the special-events department at a well-regarded, heavily patronized cultural institution in a major metropolitan area, such as the museum profiled in this case study, loses the confidence and cooperation of other institutional departments and fails to fulfill its potential for external events, it is usually a propitious time to hire an outside expert to conduct a comprehensive evaluation and assessment.

DISCOVERING THE REASONS FOR A DECLINE

Administrators at this prominent museum had concerns about their self-operated special-events department. The primary issues included its organization and relations with other internal departments, its contractual relations with outside catering firms, and its ability to increase the number of external events in accord with the capacity of the institution's spaces and staff. For us to prepare the requested assessment, it was necessary first to tour the museum's facilities, interview key staff from the special-events and other departments, and review documents including financial statements, summaries of events, promotional materials, and job descriptions. When we had completed these meetings and assessments, we listed administrators' goals and objectives for the special-events department.

As this institution had long been considered one of the premier museum and special-events venue in its city, administrators strongly expected that the special-events department should serve as a steward of the institution's resources on behalf of the community, and host event guests in a manner consistent with this role. In addition, the special-events department was charged with protecting the museum's collections and physical plant and minimizing any adverse impact on walk-in visitors resulting from its events and functions. Further, the department was expected to market its services to the community in a manner consistent with maximizing revenue for the museum while meeting the event and function needs of all internal constituencies, obviating the need for any other museum department to duplicate its services. Also charged with seeking a fair and reasonable financial return in exchange for granting caterers and other vendors the right to do business with its parent institution, the special-events department was additionally expected to apply its function policies unambiguously and consistently in regard to both internal and external events. Finally, this department was relied on to provide timely and accurate information regarding spaces' availability, rates, menus, and other critical event components.

One of the reasons the special-events department at this museum was being held to such a high set of standards was that the institution itself, besides being regarded as one of the leading cultural venues in its metropolitan area, featured public areas capable of hosting more than 1,000 guests at a time for a sit-down meal, making it a destination of choice for those seeking to book larger, more lucrative, and more prestigious functions. When working to produce events initiated by outside groups, special-events personnel were expected to respond to all telephone inquiries expeditiously, coordinate and update the master calendar of events, assign a coordinator to work with each group during pre-event planning, and arrange to provide all services, personnel, and resources necessary to stage a function that met guests' expectations. Additional duties included tracking events' status to ensure confirmation and guaranteed attendance, selling alcoholic beverages to outside caterers and managing the inventory and issuing of these products, and assigning a staff member to work with each group as its event was being staged.

Extensive as these duties were, it is important to note that special-events personnel at this museum were also expected to respond to up to 50 telephone inquiries about staging external events every day. With so much potential business arriving in the department unsolicited, it was not surprising to discover that no proactive special-events marketing program had been developed and that no promotional or sales kit had been created to attract externally sponsored events.

Given its roster of duties and heavy (though far less than optimal) event volume, this department was operating with a relatively small staff of six: a manager, an assistant manager, an administrative assistant, a facilitator, and two coordinators. The manager's job included administering the department's budget, supervising staff, selling and overseeing on-site events, and developing marketing programs to increase the museum's special-events renown and revenue. The assistant manager's primary duties included booking events and functions, helping with staff supervision, monitoring event costs, managing evening events, and aiding in the development of promotional programs. The administrative assistant was charged with maintaining the event records and master calendar, fielding incoming inquiries, and preparing post-event reports. The facilitator was primarily responsible for coordinating foodservices for internal functions, handling room setups and breakdowns, and serving as liaison to internally and externally selected caterers. For their part, the coordinators worked with event user groups to plan and execute their functions, including obtaining necessary equipment and supervising the actual staging of events. In addition, coordinators were also expected to order, receive, inventory, issue, and process payments for all alcoholic beverages, market the museum to new clients, and assist other departments seeking to hold internal special events.

UNDERSTANDING PROBLEMATIC RELATIONS WITH CATERERS AND OTHER DEPARTMENTS

We next turned our attention to tallying the number of events held and revenue earned by this museum during the past decade, as well assessing the state of its relations with its roster of caterers. Net income from special events had nearly tripled during the preceding ten years, rising to some $800,000 in the full year prior to our study and, thus, representing a significant revenue stream for the institution. Much of this increase was due to the fact that the number of events had grown by about 60 percent during this same period, while annual event guest counts had gone up nearly 100 percent; these figures were strongly influenced by the number and popularity of the special exhibits the museum had staged in different years.

More problematic for this institution was the relationship it had created between itself and local outside caterers. Not only was the museum then working with a list of 15 approved caterers (groups arranging functions here were required to select a caterer from this roster), but it was receiving *no* commissions from outside caterers for the foods and nonalcoholic beverages sold on-site, externally sponsored events. Instead, these caterers were simply paying the museum an annual fee of $600 each to maintain their listing in the exclusive group. In return for the caterers' annual $9,000 investment, they had recorded special-events sales in excess of $2.5 million during the year prior to this assessment.

Another series of special-events-related concerns at this museum was the booking, servicing, and hosting of internally sponsored functions. Contrary to common practice, many museum departments were managing their own events, expecting the special-events department only to coordinate the master calendar to ensure spaces' availability and provide alcoholic beverages when required. As a result of various museum departments' arranging their own functions through the special-events facilitator, record keeping of expenses and income was further complicated. What's more, internal special events were large and frequent affairs at this museum; nearly 100 were staged on-site in the immediate prior year, with 14 of these including more than 600 attendees. As might be expected, internal special events arranged by other museum departments were sometimes catered by firms on the special-events department's approved list but, just as often, caterers not on the list were employed at these functions.

IDENTIFYING ISSUES TO BE RECTIFIED

Based on the data we collected, we identified six major issues confronting the museum's special-events department, and offered recommendations to help resolve each. The first and perhaps overriding issue was the role of the special-events department in relation to other departments. Most cultural institutions that host as many external and internal special events as this museum did find that managing and coordinating them through a single department to be most effective. In this instance, centralizing all or almost all special events and functions under a single department would offer two important potential advantages. One was that duplication of resources and skills could be eliminated, as no other departments would have to invest in personnel capable of carrying out special-events-related duties.

We also pointed out that the museum was missing an opportunity to gain additional financial benefits from its outside special-events suppliers. If the special-events department was solely responsible for negotiating with caterers, equipment rental firms, and other vendors, the museum would be in a stronger bargaining position. What's more, if the list of approved caterers was pared down to about five, the volume of business each would be doing would then justify a sizable commission to the museum from special-events sales.

While most of the museum administrators and staff we interviewed agreed in principle with the assessment that the special-events department should be solely responsible for all functions and necessary support activities, in reality there were several more issues that needed to be addressed before such a change could be implemented. Among these was the widely shared perception that the special-events department historically had been less responsive to requests to stage internal events than for external ones. This perception was supported by reports that the department had been unwilling to coordinate such functions for other departments when the facility rental fee (which was universally applied to external events) was waived. Whether such refusals had actually occurred could not be ascertained; the key point was that members of other museum departments *believed* the special-events department to be uncooperative and often uninterested in providing requested hospitality services.

There was also a question of whether the special-events department was administratively capable of handling more functions. Doubts on that score were expressed by members of other museum departments who had had negative service experiences, including not being able to obtain timely information about the availability of event spaces, delayed invoicing, and incorrect responses to inquiries.

Further reinforcing museum personnel's concerns regarding the special-events department was this organization's adherence to several policies generally seen as unreasonable and unenforceable. Notable among these was the special-events department's reported refusal to accept or serve donated alcoholic beverages at its functions. Another complaint was that the special-events department was too slow in responding to requests from other museum departments, with one administrator recounting how a relatively simple request had received no response for an entire year. In addition, several of the departments that were then arranging and managing their own special events had, naturally enough, formed beneficial relationships with caterers and other vendors. These benefits included discounts on standard charges and other considerations, such as donations to the museum. Administrators at this cultural institution were aware that if all internal function management was shifted to the special-events department, these beneficial relationships might be jeopardized, particularly if a caterer that had been used before was not included on the approved caterer list.

WHY SPECIAL EVENTS SHOULD HAVE CENTRALIZED MANAGEMENT

In response to these concerns, as well as already identified goals and objectives, we recommended that, to the maximum extent feasible, all events and functions should be brought under the sole management of the special-events department. The potential benefits of this move would include the administration of contractual and financial relationships with all event-supporting vendors by a single, central, and presumably professionally directed organization that would be well positioned to create new economies of scale and obtain more favorable prices and commissions. It was also urged that all internal and external event-booking groups be required to employ *only* caterers on the museum's approved list. The list itself would benefit from reworking, to ensure that while fewer caterers would be included, they would be those most preferred by the institution and its guests, even if that meant excluding some formerly favored by other departments. It would therefore be essential to explain the benefits of this change to other departments' personnel, including the fact that previously received discounts and considerations were unlikely to be as financially rewarding for the museum as renegotiated exclusive relationships with a smaller group of suppliers.

In return for this heightened authority to develop special events, we recommended that the department take on a new commitment to respond fully and consistently to other departments' needs and expectations when it came time to arrange internal special events. This would obviate the need for other institutional entities to maintain relations with outside special-events suppliers or to use their staff to duplicate services available from the special-events department (though these individuals could be made primary liaisons to the events department). An additional benefit would be that, with the master events calendar fully under the control of a single office, timely information about the availability of museum event spaces would be available on a more consistent and accurate basis.

(Interestingly, a potential downside for this museum in accepting this group of recommendations was that, given the policies then in effect, the special-events department would still be facing a financial disincentive when it came to staging internal events, since these functions were not charged facility rental fees. We suggested that the special-events department could best overcome this obstacle by increasing the revenue the museum received from outside special-events vendors, as discussed in the next section.)

At some cultural institutions, it may be advisable to create an exemption for the office of the president or similar senior administrative entity that would allow its personnel to arrange small luncheons directly with outside caterers. Even in these instances, though, caterers hired to produce these small luncheons should always be drawn from those on the approved list, and should contract their work through and issue post-event reports to the special-events department.

INCREASING REVENUE
FROM OUTSIDE CATERERS

Our second set of recommendations addressed the museum's business relationships and financial arrangements with its caterers and other special-events vendors. Even though this institution was receiving no commission on catering foods and beverages sold at its special events, our experience has shown that large cultural institutions in major metropolitan areas regularly receive such commissions, which are often as high as 10 percent of total event food-and-(non-alcoholic) beverage sales. It was expected that the museum should anticipate receiving commissions from all caterers on its approved list, but that the list should be cut down (from fifteen approved vendors to five) and that the practice of hiring catering firms not on the list be eliminated. Taking these steps alone, it was estimated, would likely give this museum better than $200,000 a year in new special-events net income.

In addition, to help determine which caterers should be on the reduced list and to gain the most complete picture possible of how personnel rated all the caterers now serving their special events (and the amount of annual revenue they were generating at the museum), institutional personnel were surveyed. Input was also specifically solicited on past caterers from members of the security, housekeeping, and visitor-services departments. These interviews revealed that certain caterers had attempted to evade paying the fee required to use the museum's trash compactor and had placed trash in the institution's garbage receptacles without paying. Other concerns were catering staff who entered the museum through nonprescribed entrances and changed into uniforms in public restrooms, and the fact that caterers' staff were not required to pass through security inspections as they left the museum, making it impossible to determine if any were carrying out stolen goods.

With all this information in hand, we advised this museum's special-events department to begin a request-for-proposal (RFP) process in order to identify and select approximately five local caterers that would be given exclusive rights to handle on-site functions. In order to provide the best possible outcome, an RFP should require caterers to identify the commissions they would pay on on-site food sales (alcoholic beverages were sold by the museum) and a guaranteed minimum payment due the institution if commissions failed to reach agreed-upon levels. While it is uncommon for outside caterers to guarantee annual minimum payments to cultural institutions, this museum had earned sufficient local prominence from its annual special-events volume, and stature to impose this additional requirement on its suppliers.

In addition to asking caterers to propose "guaranteed" revenue they would bring to the museum annually, it was recommended that they also cite the penalty they would incur if guaranteed sales were not achieved. Further, the RFP would help to eliminate other existing problems if it proposed a consistent 20 percent discount from caterers' standard prices for all internal events generated by the museum's departments and eliminated caterer standard service charges, which typically range from 15%–18% of the total catering billing. To ensure buy-in by other institutional departments, and those whose personnel made use of catered events, they would be given a say in which caterers would be included on the revised approved list. We suggested that the museum would be best served if it considered caterers that not only offered diverse menus, services, prices, capacity specialization, and event expertise, but also agreed to work under a single, uniform contract of terms and conditions drawn up by the institution.

Two more recommendations were also proposed, to maximize any new agreement's effectiveness. The first was that the museum needed to do a more stringent job of enforcing its policies vis-à-vis caterers' personnel, including reporting all violations in writing to the special-events department and enforcing a three-strikes-and-you're-out rule (or other penalties appropriate to the severity of the violations) against those personnel who failed to adhere to these policies. It seemed only sensible to couple this advice with a suggestion that museum security begin checking all catering staff as they left the facilities, including inspecting bags, backpacks, and coats to ascertain if any contained stolen or unauthorized property.

The final idea put forth to address this museum's business relationship issues was that a similar approach be followed with such other suppliers as equipment rental firms, electrical services providers, florists, decorators, and photographers. To the greatest extent feasible, vendors of each sort should be qualified and added to an approved list, and agreements should be developed with these listed vendors defining the commissions and/or rebates they would pay to the museum.

IMPROVING SPECIAL-EVENTS MANAGEMENT PERFORMANCE

The third issue for which recommendations were developed was the internal organization and operation of the special-events department itself. Because input on the department had already been gathered from members of other institutional entities, we recorded the perceptions of special-events personnel about their own performance and studied this information before offering an assessment. For example, special-events department staff told us that it took them time to respond to requests for information because calendars and reservation books were still kept by hand, rather than on computer, meaning that information on bookings' status was not always up-to-date or completely accurate. Because internal events were not always booked through the department, space conflicts had in fact occurred. And, despite several years of development with an outside supplier, a custom event-management software package was still not ready for implementation. Added to these shortcomings was the fact that policies, especially those pertaining to pricing, had not been administered consistently, creating "gray areas" in operations that different special-events staff interpreted differently.

Of particular importance (given its financial significance) was the special-events staff's attitude toward the management of alcohol operations and sales. This task was generally perceived as a burden by event coordinators, who reported that they were regularly taken away from client-service duties to receive liquor shipments at the loading dock, check beverages out to event groups, or place re-stocking orders after functions.

We had noted that the department's manager was positioned in such a way that every special-events staff member was considered to be a direct report. This structure (insofar as it indicated actual practices) all but ensured that the manager's time was overly taken up by daily coordination issues at the expense of long-term de-partmental planning. Therefore, it was unsurprising that special-events staff felt the department tended to work in a reactive rather than a proactive mode, and that there was an absence of teamwork between colleagues. Employees here were also concerned that staff members who initiated event planning with a particular group were not always assigned to work with that group during their function. This policy had added to coordination difficulties and did not seem to be desirable from clients' perspective.

In sum, it seemed that this museum's special-events department was set up to handle daily management of external events with rel-atively little difficulty. However, its organization left it ill-equipped to assume responsibility for management of internal events and functions, since appropriate systems were not in place, policies and practices had become confused, morale was low, and trust (both within the department and between it and other departments) was far weaker than it needed to be.

Given this situation, a package of recommendations was offered. To begin the department's improvements, its manager should be encouraged to focus on long-term development of the department by winning back the confidence of other departments' personnel, revising policies and practicing them consistently, developing and implementing a strategic marketing plan, and installing event-management software that allowed real-time checks on the status of all prospective and confirmed events. The department's assistant manager would then be able to concentrate on overseeing daily op-erations, including working to improve staff morale and teamwork, coordination of events (by minimizing handoffs of responsibilities), and vendor compliance with museum standards and policies.

Event coordinators could best contribute by concentrating on helping event groups plan their functions and by marketing and promoting the museum to create new business. It was advised that a new position be added to the special-events staff—that of marketing coordinator, who would be expected to develop new business with private caterers, event planners, destination bookers, and other demand generators. The marketing coordinator would also help generate new business leads in support of event coordinators' efforts and assist in the creation of promotional materials and sales kits.

It was also advised that event coordinators could perform client-related functions best if they were no longer accountable for the purchasing, receiving, and storage of alcoholic beverages; however, it seemed appropriate that they continue to handle issuing and post-event restocking responsibilities. Alcohol-related duties formerly performed by the coordinators could become the obligation of the facilitator, who would also coordinate smaller internal events. It should be noted that the sale of alcoholic beverages had accounted for over 45 percent of this museum's special-events total revenue in the year before this assessment, earning nearly $700,000 in new revenue after an investment of under $150,000 in the cost of goods sold. With so much income at stake, responsibility for the management of alcoholic beverage sales should never be ambiguous, regardless of how burdensome these duties are perceived to be.

ENHANCING MARKETING PLANS AND STRATEGY

The fourth issue addressed in our assessment was the marketing plan and growth strategy in place for this institution's special-events department. While growth was strong, as evidenced by the steady stream of telephone requests for special events, it also fostered an "order-taking" mentality among staff, which discouraged both the development of a strategic plan and the creation of suitable marketing materials. We also pointed out that for all its volume, the mix of business arriving from telephone inquiries was not likely the ideal variety of events for the museum. The lack of a suitably sophisticated promotional kit was further seen as putting this museum at a competitive disadvantage in regard to attracting high-end and larger functions compared with other cultural institutions in the area.

With growth and future success in mind, we therefore recommended that the special-events department draft a strategic marketing plan that included a description of the qualities that favorably distinguished this museum from other local museums, casting it as the premier local cultural venue. Such a plan should also determine primary market segments and demand generators, set up a pricing strategy that took advantage of the museum's local stature, and identify promotional activities (including advertising) necessary to generate growth in event bookings in accord with administrators' expectations. The final component of this plan should be a staffing and resource assessment that would determine the level of human resources required to support different growth targets, not only for the special-events department, but also for the housekeeping and security departments.

We next outlined the recommended contents for the promotional kit: a personal letter to each recipient, color photos of events held in different museum spaces, information on upcoming exhibits, rental fees and capacities for all event areas, the approved list of caterers, a summary of steps involved in planning functions, the endorsements of prior comparable clients, and a response card to be returned to the special-events department. These materials and the kit as a whole would probably be most successful if they were designed by an outside company that specialized in developing identity and marketing materials for cultural institutions.

ASSESSING POTENTIAL FOR REVENUE GROWTH

The fifth issue for which recommendations were provided was the potential for revenue growth from special events. Based on our assessment, three distinct opportunities for growth were apparent: increasing the number of externally-hosted special events; enhancing events' profitability; and generating more revenue by improving contractual arrangements with caterers and other vendors.

Increasing the number of events staged annually at this museum could be achieved, in part, through greater marketing to and capture of appropriately high-end civic groups and corporations. However, it is important to note that a cultural institution's ability to increase special-events totals in addition depends upon its capacity. In the case of this museum, the most recent annual usage of its largest space was no greater than 35 percent of its potential. While our study could not account for all (internal) events, nor holidays or seasonal fluctuations, it clearly indicated that this largest space offered sufficient capacity for meaningful growth in the number of on-site special events. For example, we calculated that if the museum increased special-events guest totals by just 10 percent over the level achieved in the most recent year, it would realize new gross revenue of $200,000 and net profits of nearly $90,000.

To help this institution increase the profitability of its special events, we recommended reviewing the rates charged to external user groups and encouraging the department to book more large events with comparatively high food-and-beverage selling prices. We pointed out that if the museum could realize just *one* dollar more in revenue per event guest above the $20.50 it had recorded in the prior year, net income would increase by some $90,000. The impact of the third opportunity, deriving commissions in exchange for granting caterers and other vendors exclusive rights to do business at the museum, could also be calculated in financial terms. For example, if this cultural institution instituted a 10 percent commission structure on event catering and 5 percent on equipment rentals, new annual revenue would total over $280,000.

We concluded that the potential for increased special-events revenue warranted the investment the museum would have to make in new marketing materials and an additional department member dedicated to marketing functions to outside groups, as well as support staff who could be brought on board in accord with actual revenue growth.

WHEN SHOULD SPECIAL EVENTS BE OUTSOURCED?

Our final recommendations concerned the pros and cons of outsourcing the management of special events. Given the number of challenges facing this special-events department, it seemed imperative to consider whether an outside foodservice provider should replace the self-managed organization. Because challenges of this sort often confront self-operated special-events programs in cultural institutions, it's worth discussing them again. First, the special-events department had to regain the confidence of other departments while assuming esponsibility for all museum-based functions, both internal and external. Other requirements included reorganizing the department's structure to make it more responsive to internal clients, recruiting new staff, installing an event-management information system to provide real-time access to event status and space availability, and putting into action a strategic marketing plan that would maximize special events' contribution to the institution's general fund.

Based on these assessed factors, it appeared that the museum could choose among three approaches to meet special-events challenges: working with current department personnel to create necessary changes, replacing special-events personnel as part of a departmental reorganization and giving new hires the task of making these improvements, or contracting with an outside organization to manage the museum's special events and successfully resolve current challenges. Each approach needed to be comparatively evaluated for advantages and disadvantages before administrators could decide which option would best serve the museum.

The first approach, retaining current special-events management, promised to minimize consternation among employees and loss of important institutional knowledge about how things were done at this museum. Its potential drawbacks included the likelihood of resistance to change, difficulty in re-establishing rapport between existing employees, and doubts about the current manager's ability to implement needed improvements.

The museum's second option, hiring new personnel to implement improvements, offered the institution a chance to recruit directly for specific competencies and experiences. It was also probable that new hires would effect changes at a more rapid pace. Potential downsides, however, included the necessity of transferring or terminating for cause several current staff and the resulting negative impact on morale among remaining special-events employees. Also to be considered were the possibility that new hires might require greater compensation than incumbents, the inevitable loss of accumulated institutional knowledge, and the likelihood that the department's capability would improve only as quickly as new staff were hired and brought up to speed.

The third choice, outsourcing this department to a foodservice management company, would likely provide greater special-events expertise from day one, allow administrators to oversee the functions program from a distance and concentrate their energies on fulfilling the museum's core mission, and make individual staffing decisions more flexible, as personnel could be changed without replacing the operator. Mitigating against this choice was the acknowledgment that any outside firm would need time to learn how the museum functioned, and the potential conflict of interest that would arise if the selected operator competed against other firms on the approved list for catering. In addition, the available number of qualified special-events operators was likely to be small, new relationships would have to be built immediately rather than over time, and the outside company would inherit none of the goodwill and all of the problems created by the self-operated department.

After considering all these factors, we recommended that this cultural institution consider the outsourcing option for its special-events program. To begin the process, we suggested that all museum administrators whose areas of responsibility would be affected by a decision to outsource meet to identify and address all internal issues likely to result from the process. We also proposed engaging in informal, confidential discussions with prospective commercial-sector operators. There would be two purposes to these meetings: to assess these operators' experience and ability to address the issues then facing this special-events department; and to determine whether a sufficient number of firms would be likely to respond to the proposed RFP to ensure a successful outcome to the process.

Last, if administrators found both that outsourcing-related internal issues could be effectively dealt with and that a sufficient number of qualified and interested potential new operators could be found, the museum could begin to solicit proposals for special-events management. If *either* criterion could not be met, however, we advised that the other approaches be reconsidered. It was further stressed that the museum should not make known its intention to prepare a RFP unless and until its senior staff had firmly committed to the new operator-selection process.

The museum eventually decided to proceed with our second option, hiring a new special-events department head and maintaining the entire operation in-house. As a result of the implementation of a new short list of approved caterers and other recommendations detailed in this case study, the museum's total special-events net revenue has grown about 25 percent.

Analysis of Demand and Architectural Program Statement for Foodservice Planning in a Museum Building

We were called upon by an established medium-sized museum that was undergoing a major expansion and adding a new wing to its building to provide an architectural planning statement to guide the design of one or more new dining facilities within the new and expanded building. We used (and recommend) the following approach in estimating the demand for foodservice.

To begin, the museum provided daily visitor counts for a one-year period, including total admissions, as well as an estimate of how many children and how many adults had attended.* (Children's admissions reflected the number of children arriving with school groups, rather than children visiting the museum with their families.)

We used daily visitor counts to prepare our estimate of the demand for foodservice because they captured weekly and seasonal attendance patterns. Weekends typically had more visitors than

---

*The museum was then tracking school group attendance on a monthly basis rather than daily. Daily estimates of children's attendance were computed by dividing the monthly total by the number of weekdays the museum was open (Tuesdays through Fridays) in the month.

weekdays, and warm months brought more visitors than cold months. These patterns were important, as they helped us determine the size of the foodservice operations necessary to meet demand.

We multiplied the daily adult admissions by a factor of 2.75 to estimate the attendance in the main building once the renovation and expansion were complete. We multiplied the daily children's attendance by a factor of 1.25 to estimate the attendance of school groups once the renovation and expansion were complete. The lower factor for school groups was based on the assumption that increases in group admissions are more difficult to achieve than increases in individual admissions. These growth factors were derived by calculating how much current visitation would need to grow to reach this institution's goal of attracting a total of approximately 400,000 persons annually (including school groups) in the main museum building.

We estimated a foodservice capture rate of 35 percent for adult admissions; that is, we projected that 35 percent of adult visitors would purchase lunch in the main building. Our 35 percent capture rate assumed that the planned foodservice operation would be centrally located in an admission free area where visitors would pass it both entering and leaving the museum. If the foodservice operation had been located in a less central location or a location where an admission had to be paid to gain access, we would have used a capture rate of 20 percent or less.

The museum was considering whether to set aside an area where school groups could congregate for meals. We assumed that those students would bring sack lunches but would need vending machines to provide supplements such as snacks, drinks, fruit, and similar fare.

Using the approach described above, we also projected foodservice demand based on estimated adult admissions. We graphed projected demand from month to month for a year, with separate lines for each day of the week (Figure 8.1).

An examination of current attendance figures revealed that we should project higher foodservice demands for weekends (Saturdays and Sundays) than for weekdays. To isolate this difference, we created two additional graphs. The first graph (Figure 8.2) plots projected demand for weekend days for 53 weeks; the second graph (Figure 8.3) plots projected demand for weekdays during the same period.

(text continued on p. 161)

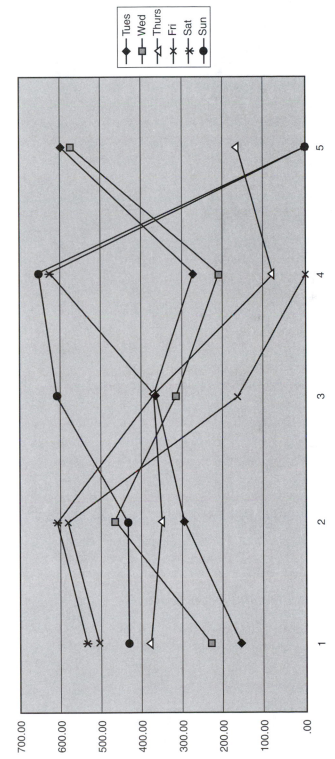

January

Figure 8.1 Projected foodservice demand by month for one year.

February

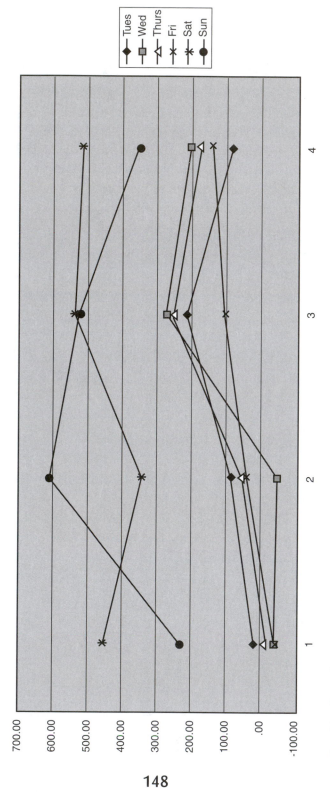

Legend:
- Tues
- Wed
- Thurs
- Fri
- Sat
- Sun

Figure 8.1 (Continued)

148

March

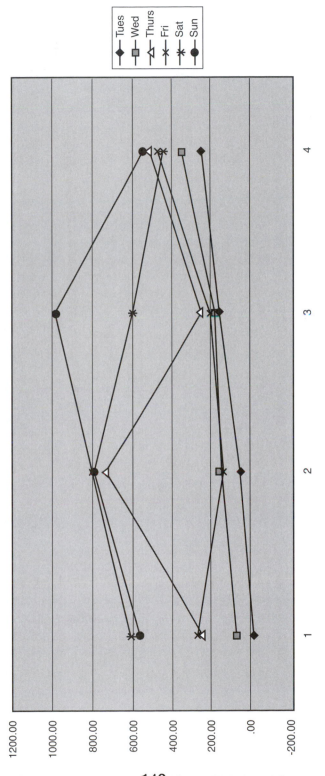

149

Figure 8.1 (Continued)

April

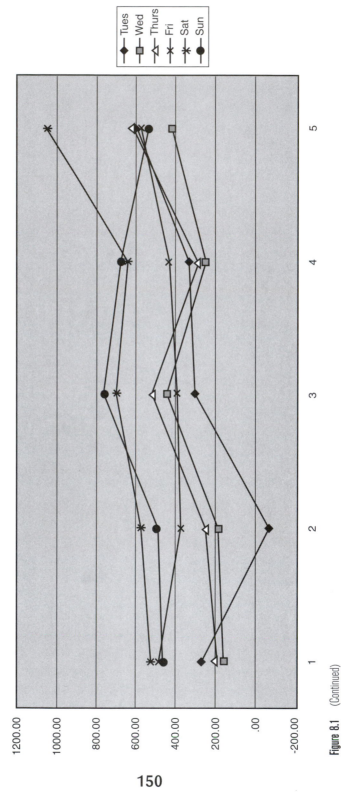

Figure 8.1 (Continued)

May

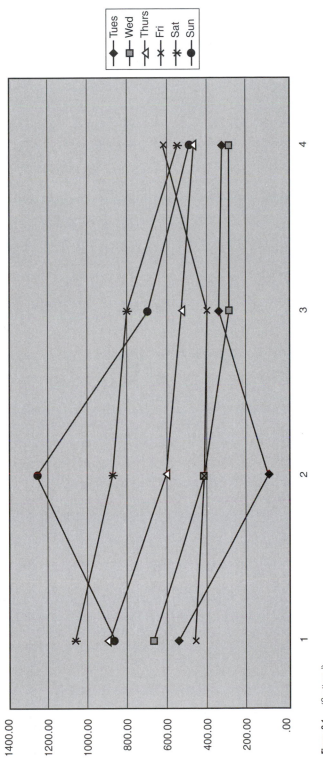

Figure 8.1 (Continued)

June

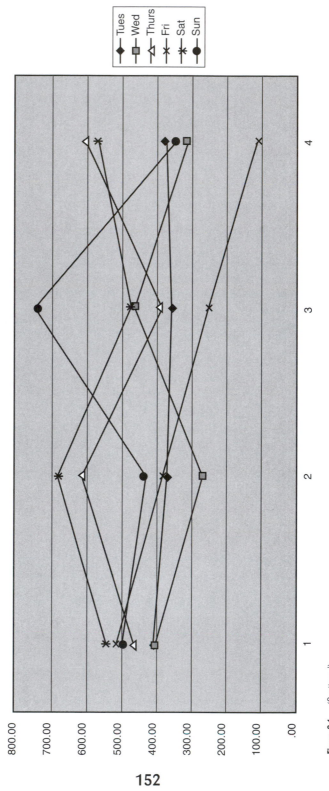

Figure 8.1 (Continued)

July

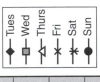

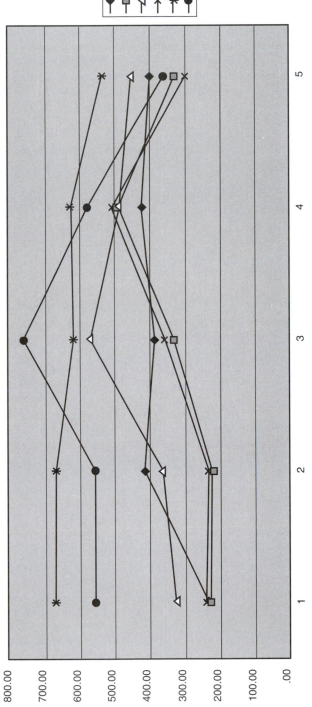

Figure 8.1 (Continued)

153

August

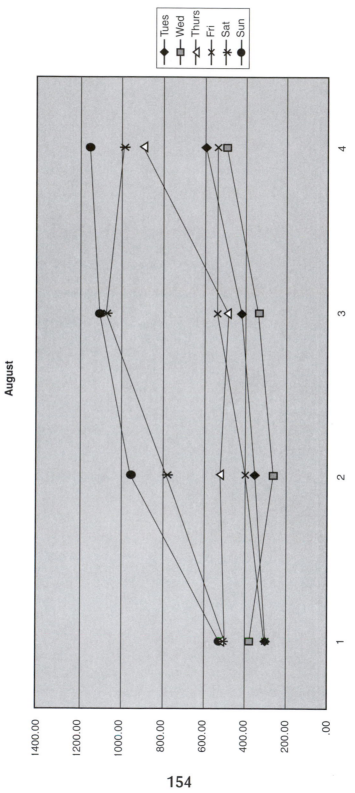

Legend:
- Tues
- Wed
- Thurs
- Fri
- Sat
- Sun

1400.00
1200.00
1000.00
800.00
600.00
400.00
200.00
.00

1   2   3   4

154

Figure 8.1 (Continued)

September

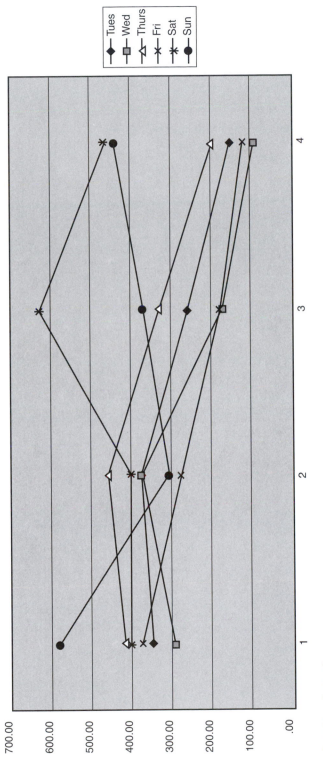

Figure 8.1 (Continued)

October

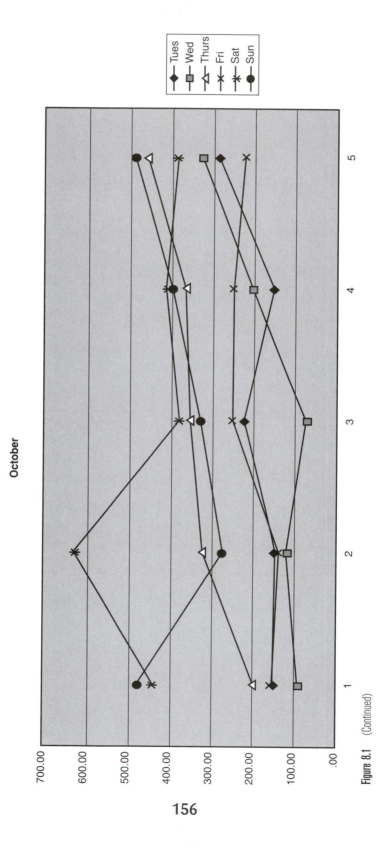

Figure 8.1 (Continued)

November

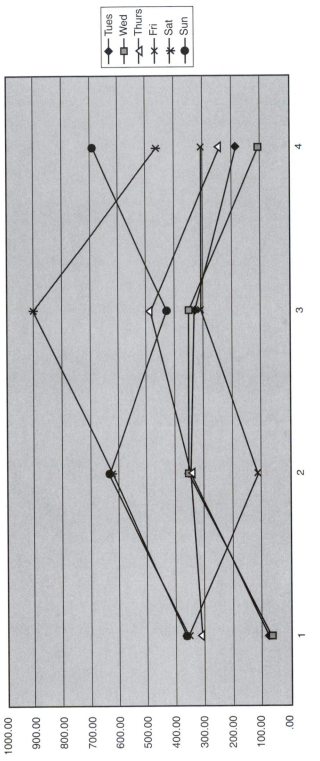

Figure 8.1 (Continued)

December

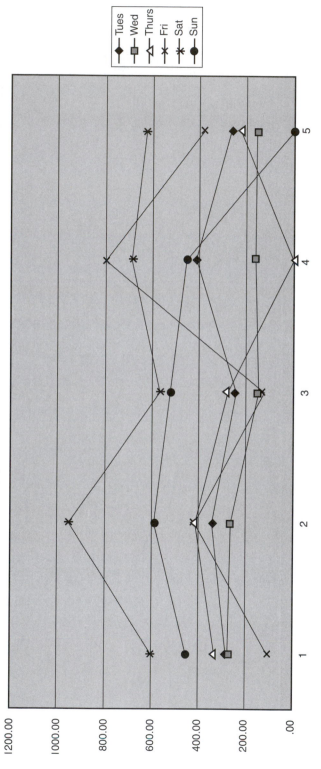

Figure 8.1 (Continued)

158

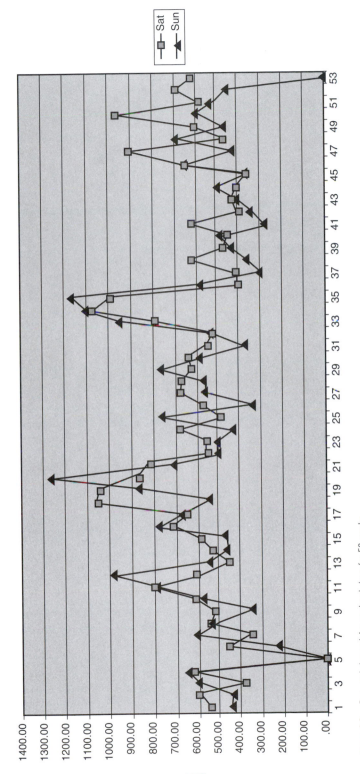

Figure 8.2 Projected demand for weekend days for 53 weeks.

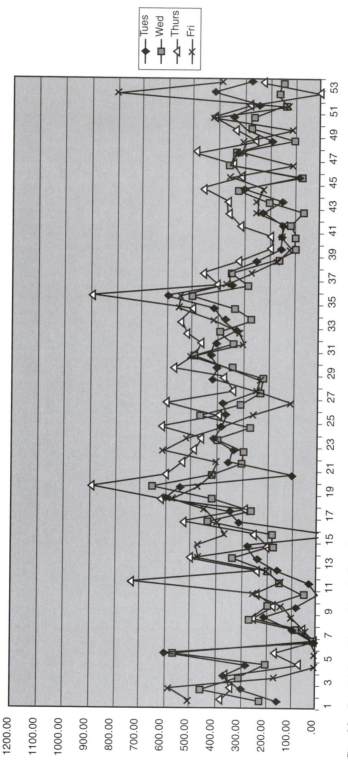

Figure 8.3 Projected demand for weekdays for 53 weeks.

160

An inspection of these graphs of projected demand suggests the following:

- A foodservice facility capable of serving approximately 600 people per day would meet the demand for foodservice on all but a few weekdays.

- However, a foodservice capacity of 600 would be sufficient only on about 50 percent of weekend days. A total capacity of approximately 900 people per day, though, would be adequate on all but the busiest weekend days.

MATCHING FACILITY DESIGN TO DEMAND

One approach to meeting the projected foodservice demand would be to design and build a single facility with a capacity of 900 people per day. However, that approach has several distinct disadvantages:

- A foodservice facility designed for 900 people per day would seem empty on most weekdays—particularly during the colder months, when demand projections forecast a volume of about 350 people per day.

- The foodservice operator would need to staff, clean, and maintain a facility that would be substantially larger than needed on most days it was open, resulting in higher-than-optimal operating costs.

An alternative way of meeting the projected demand would be to have two foodservice facilities. One facility would have a capacity of 600 people per day and be open every day of the year. The second foodservice facility would have a capacity of 300 people per day and would be open primarily on weekends during warm months and at other times as demand warranted. The physical space of the second foodservice facility would double as a special-events and catering venue. There are a couple of advantages to this approach:

- The foodservice operator would have greater control over expenses by being able to open the second facility only when demand warranted.

> ■ A second facility would create greater flexibility in foodservice offerings. For example, the facility could be set up for cart service on most days, but provide table service for special events.

MEASURING SCHOOL GROUPS

The average number of children in daily attendance was approximately 160. Inasmuch as the area dedicated for school groups could be used multiple times during a day (for example, one school group could be scheduled for 11:00–11:45 and a second for 12:00–12:45), a space with a seating capacity of 100 persons appeared to be adequate.

Although schoolchildren could purchase their lunches in the same facility as adult visitors, that approach could pose supervision problems for school group leaders. Also, large groups of children going through a main foodservice facility are generally difficult to serve efficiently. An alternative would be to recommend that school groups bring lunches from home and/or to provide vending service (milk, snacks, juice, soda, fruit) to supplement these lunches.

RECOMMENDATIONS REGARDING OVERALL APPROACH TO FOODSERVICE IN THE MAIN MUSEUM BUILDING

Because of the seasonal and weekly attendance fluctuations in projected demand at this museum, we presented the following recommendations:

> ■ A primary foodservice facility with a capacity of 600 persons per day (and containing 250 seats) should be located on the lower level, adjacent to the central visitor circulation spine of the main building. This facility would be open every day that the main museum building was open. When administrators are determining the size of (or how many square feet should be devoted to) a new foodservice, past, current, and future visitor counts should be reviewed by day of week and

month, ideally over a 12-month period. The restaurant should not be sized to accommodate the largest potential number of guests but, rather, for the average peak visitor count. After the average peak attendance number is determined, the square footage for the new facility can be planned based on the following rule of thumb (assuming an average peak visitor day of 900):

- Apply the anticipated capture rate (20% to 35%).

- Assume the dining room's seating space will accommodate a minimum of two to three turnovers with an 85% seat occupancy rate during an average lunch period.

- Example: 900 × 30% Capture Rate = 270 customers

- A secondary (overflow) dining facility with a capacity of 300 persons should be located on the museum's plaza level. This facility would be open on weekends to serve lunch. During weekdays and evenings, it would be available as a site for special events. It would have a staging kitchen, rather than a full kitchen. Food would be presented in attractively designed carts that could be removed when the space was needed for other purposes. However, table service would also be an option.

- The school-group room should have a capacity of 100 people, as well as space for user traffic flow and vending machines. Furnishings should be portable, cafeteria-style tables with attached benches.

ARCHITECTURAL PROGRAM GUIDELINES FOR FOODSERVICE SPACES

The following pages provide architectural program information based on these recommendations. This document's importance lies in the precise and detailed requirements it provides to facility and space designers. A similar document should be created whenever a foodservice facility is being planned or prepared for renovation. The architectural program statement should be organized according to primary functional areas.

Name of Area: Receiving

Description of Function	Food products are unloaded from trucks at the dock and brought to the receiving area, where they are inspected, weighed, and taken to dry or refrigerated storage.
	Foods prepared for catered events in other areas of the museum depart from the receiving area.
	Garbage is stored in a Dumpster (or compactor) at the dock.
Relationship to Other Areas	Access to truck delivery
	Adjacent to dry and refrigerated storage on same floor or accessible by elevator
	Adjacent to food waste/trash holding area
Net Square Footage Required	100 (does not include the dock itself or truck parking)

Finishes		
	Ceiling	Washable, such as a dropped ceiling with cleanable tiles
	Walls	Resistant to damage from hand trucks, carts, and other food transport equipment (e.g., epoxy-sealed concrete block, structural glazed tile)
	Floor	Quarry tile or other material that is impervious to moisture and easily cleaned and sanitized

Types of Equipment	Receiving table and scale
Special Design Considerations	Exterior door at least 42 inches wide to accommodate hand trucks
	Air curtain installed above door to prevent entry of insects and other pests
	Lockable door with doorbell to summon receiving personnel
	Control of odors from Dumpster

Name of Area: Bulk Dry Storage

Description of Function	Dry goods (canned, packaged), paper products, and other items are stored on shelving or dunnage racks prior to use.
Relationship to Other Areas	Adjacent to receiving area Adjacent to pre-preparation area
Net Square Footage Required	250

Finishes

	Ceiling	Washable, such as a dropped ceiling with cleanable tiles
	Walls	Resistant to damage from hand trucks, carts, and other food transport equipment (e.g., epoxy-sealed concrete block, structural glazed tile)
	Floor	Quarry tile, sealed concrete, or other material that is impervious to moisture and easily cleaned and sanitized

Types of Equipment	Shelving and dunnage racks
Special Design Considerations	Minimum useful clear width for a dry-storage area: 8 feet Irregularly shaped (nonrectangular) spaces to be avoided to maximize space utilization Minimum door opening 36 inches; 42 inches is preferred Lockable door to secure access

Name of Area: Bulk Cold Storage

Description of Function	Refrigerated and frozen products are stored on fixed shelving and dunnage racks prior to use in the foodservice operation.
	Products that have been prepared (e.g. salads, desserts) and stored on mobile racks prior to service.
Relationship to Other Areas	Adjacent to receiving area
	Adjacent to pre-preparation area
Net Square Footage Required	250 (125 each for cooler and freezer)

Finishes	*Ceiling*	Manufacturer's prefabricated insulated panel, interior painted white
	Walls	Manufacturer's prefabricated insulated panel, interior and exposed exterior painted white
	Floor	Manufacturer's prefabricated insulated panel; quarry tile over panel, or integral diamond-tread plate-floor system

Types of Equipment	Refrigeration systems and prefabricated insulated panels
	Shelving, dunngea racks, and mobile racks
Special Design Considerations	In new construction, walk-ins to be erected in a 4-inch deep pit in an unfinished floor and quarry tile installed to run continuously from outside to inside
	If set up over existing floors, walk-ins to be provided with integral diamond-tread plate-floor panels and interior ramps
	Remote refrigeration systems linked from outdoor locations (rooftop or external concrete pad)
	Minimum effective width of walk-in units: 8 feet
	Irregularly shaped spaces do not accommodate prefabricated walk-in units cost-effectively

Name of Area: Pre-preparation

Description of Function	Washing, trimming, cutting, mixing, slicing, and other preparation processes are applied to foods prior to cooking. Salads and other noncooked items are also prepared in this area.
Relationship to Other Areas	Adjacent to walk-in coolers and freezers and dry storage Adjacent to final preparation area
Net Square Footage Required	400
Finishes	*Ceiling* Washable, such as a dropped ceiling with cleanable tiles *Walls* Resistant to damage from hand trucks, carts, and other food transport equipment (e.g., epoxy-sealed concrete block, structural glazed tile) *Floor* Quarry tile or other material that is impervious to moisture and easily cleaned and sanitized
Types of Equipment	Work counters and tables with sinks; hand sink(s) Disposer for food waste Mixers, choppers, slicers, and other processing equipment
Special Design Considerations	None

Name of Area: Final Preparation

Description of	Food is cooked prior to service.
	Food is garnished for service.
Relationship to Other Areas	Adjacent to pre-preparation area
	Adjacent to servery
Net Square Footage Required	600

Finishes

Ceiling	Washable, such as a dropped ceiling with cleanable tiles
Walls	Resistant to damage from hand trucks, carts, and other food transport equipment (e.g., epoxy-sealed concrete block, structural glazed tile)
Floor	Quarry tile or other material that is impervious to moisture and easily cleaned and sanitized

Types of Equipment

Cooking equipment, including oven, kettles, steamers, charbroilers, ranges, and other pieces as deemed appropriate to the menu; hand sink(s)

Worktables for assembly and garnishing of food items

Reach-in refrigerators and freezers for cold-holding of products used daily

Kitchen ventilation systems with fire suppression apparatus

Special Design Considerations

Equipment's intensive gas, electric, and water requirements may be most economically met by a utility distribution system.

Ease of cleaning is facilitated by mounting equipment on casters.

Kitchen ventilation requires integral supply of air tempered to 55°F.

Ventilation fire suppression systems can be integrated with building sprinkler systems, code permitting.

Name of Area: Servery–Self-Service

Description of Function	Holds and merchandises all menu items available to purchase
Relationship to Other Areas	Adjacent to final preparation area Adjacent to seating area
Net Square Footage Required	1,400

Finishes		
	Ceiling	Dropped
	Walls	Ceramic tile
	Floor	Tile, pavers, or other materials that are impervious to moisture and easily cleaned and sanitized

Types of Equipment	Self-service and/or employee-service display counters at each point of service, including beverages, salads, soups, different types of entrees, desserts, and other items as required by menus
	Dispensing equipment as determined by menus, such as ice and beverage dispensers, condiment dispensers, and similar equipment
	Holding units behind counters, such as refrigerators and hot food holding carts; hand sink(s)
	Final preparation equipment, such as pizza ovens or grills, may be located in the servery to create a "cooked-fresh" display preparation theme

Special Design Considerations	Capacity of the facility is determined by the rate at which visitors can flow through the servery, which is a function of serving-station layout (straight-line versus scatter) and the number of customer service points.
	Display-counter design, lighting, and treatments should harmonize to create an inviting atmosphere in which foods and beverages are shown to their best advantage.
	Some final preparation processes may occur at points of service (e.g., display preparation) and therefore require ventilation and fire-suppression systems.

Name of Area: Dining–Café

Description of Function	Visitors eat at tables in the dining area.
Relationship to Other Areas	Adjacent to servery Adjacent to bussing point
Net Square Footage Required	2,520 (to accommodate 180 seats and traffic circulation)
Finishes	*Ceiling* As appropriate to environment *Walls* As appropriate to environment *Floor* Easily cleaned and maintained
Types of Equipment	Table, chairs, bus stations for trays, high chairs for children
Special Design Considerations	Mix of two-top, four-top, and larger tables should be appropriate to size and composition of visitor parties. Tables and chairs should be easy to clean.

Name of Area: Ware Washing

Description of Function	Cleaning and sanitizing of dishes, flatware, trays, glasses, cups, utensils, pots, and pans after use
Relationship to Other Areas	Adjacent to dining area and servery
Net Square Footage Required	250

Finishes		
	Ceiling	Washable, such as a dropped ceiling with cleanable tiles
	Walls	Resistant to damage from hand trucks, carts, and other food transport equipment (e.g., epoxy-sealed concrete block, structural glazed tile)
	Floor	Quarry tile or other material that is impervious to moisture and easily cleaned and sanitized

Types of Equipment	Tables for clean and soiled dishes, disposer, dishwashing machine, hot water booster heater, pot sink, dish and plate storage carts, and garbage receptacles; hand sink(s)
Special Design Considerations	Hot water supply of 140°F, capable of being raised to 180°F by booster heater.
	Ventilation (exhaust only) required for dishwashing machine.
	Additional room ventilation required to control humidity and moisture.
	Secure storage for cleaning supplies required.

Name of Area: Foodservice Management Offices

Description of Function	Work areas for foodservice management personnel
Relationship to Other Areas	Adjacent to pre-preparation and final preparation areas, with view-through window(s)
Net Square Footage Required	180
Finishes	***Ceiling*** Dropped ceiling ***Walls*** ***Floor*** Quarry tile or other material that is impervious to moisture and easily cleaned and sanitized
Types of Equipment	Desk, chair, file cabinet, computer, printer, copier, telephone
Special Design Considerations	Windows allowing view of pre-preparation and final preparation areas Phone connections, multiple data lines, dedicated circuits for computer systems

Name of Area: Employee Rest Rooms/Locker Room

Description of Function	Foodservice employees store personal belongings in locked cabinets, change into uniforms, and attend to personal hygiene needs.
Relationship to Other Areas	Adjacent to employee entrance; shielded from public view
Net Square Footage Required	500

Finishes	*Ceiling*	Washable, such as a dropped ceiling with cleanable tiles
	Walls	Easily cleaned and sanitized; ceramic tile, epoxy-sealed concrete block, or fiberglass-reinforced panel
	Floor	Quarry tile or other material that is impervious to moisture and easily cleaned and sanitized

Types of Equipment	Lockers
	WC

Special Design Considerations	None

Name of Area: Staging Kitchen for Special Events/ Secondary Dining

Description of Function	Bulk foods brought from main kitchen are plated and served. Plates are cleaned and scraped, then returned to main dishroom.
Relationship to Other Areas	Adjacent to table-service seating area Service corridors and elevator (if necessary) to main kitchen
Net Square Footage Required	420
Finishes	*Ceiling* — Washable, such as a dropped ceiling with cleanable tiles *Walls* — Resistant to damage from hand trucks, carts, and other food transport equipment (e.g., epoxy-sealed concrete block, structural glazed tile) *Floor* — Quarry tile or other material that is impervious to moisture and easily cleaned and sanitized
Types of Equipment	Hot and cold food holding units; counter for plating; beverage dispensers; bus carts; trash accumulation
Special Design Considerations	Easy access to dining area for servers moving in both directions Minimum 36-inch-wide doors to allow tray service

Name of Area: Secondary Dining/Special Events

Description of Function	Visitors eat at tables in the dining area. Table service with waitstaff
Relationship to Other Areas	Adjacent to staging kitchen
Net Square Footage Required	1,675 (90 seats plus service and circulation areas), used only for overflow dining
Finishes	*Ceiling* As appropriate to environment *Walls* As appropriate to environment *Floor* Carpet
Types of Equipment	Table, chairs, high chairs Carts
Special Design Considerations	Mix of two-top, four-top, and larger tables as appropriate to size of visitor parties Note: It would be desirable to have space available for 250 seats (3,250 net square feet) to accommodate larger special events. If so, the ability to divide the room into two separate spaces as needed also would be desirable.

Name of Area: Catering Storage

Description of Function	Accommodates the storage of catering equipment used to provide foodservice within and outside of the main museum building
Relationship to Other Areas	Adjacent to primary catering spaces if most catering is in the main building Adjacent to loading dock if most catering occurs at locations outside of the main building
Net Square Footage Required	400
Finishes	**Ceiling** Washable, such as a dropped ceiling with cleanable tiles **Walls** Resistant to damage from hand trucks, carts, and other food transport equipment (e.g., sealed concrete block, structural glazed tile) **Floor** Quarry tile or other material that is impervious to moisture and easily cleaned and sanitized
Types of Equipment	Shelving, banquet carts, tables and chairs
Special Design Considerations	Minimum 42-inch-wide door openings; double 30-inch-wide doors without center jamb preferred

Name of Area: School Group Room

Description of Function	Provides space for school groups to congregate and eat. Intended for use by multiple groups during a single day. Vending machines.
Relationship to Other Areas	Service access to kitchen
Net Square Footage Required	1,500 (includes 300 square feet for vending and traffic circulation)
Finishes	*Ceiling* As appropriate to environment *Walls* As appropriate to environment *Floor* Institutional-grade vinyl floor covering or other easily cleaned material
Types of Equipment	Cafeteria seating Vending machines (milk, juice, canned sodas, fruit, snacks)
Special Design Considerations	Portable, cafeteria-style tables with integrated benches

9

Conducting Market Research for Restaurants and Special Events

If administrators want to know whether the foodservice or catering operation at their museum, aquarium, science center, zoo, botanic garden, or other cultural institution is squarely hitting the mark with target audience groups, they will need specific information to make that determination. And the way to uncover the kind of information administrators need in order to maximize the sales potential of foodservice and special-events functions is to go straight to the source—current visitors, customers, potential future customers, and even competitors.

Market research can provide an invaluable tool for discovering this information, describing and defining the dining likes, dislikes, and preferences of members of target markets. Professional market research can provide those responsible for foodservice and catering programs with solid data that can be analyzed and reviewed in order to make viable suggestions for future program changes. Findings gleaned from surveys and focus groups give administrators and operators reliable information about who their customers are and what they want and expect, thus helping foodservice and

special-events operations succeed. Administrators who are dissatisfied with their current restaurant or special-events sales can expect
the information gained from market research to provide answers
that will steer them toward better performance and increased profits. Even if administrators are not dissatisfied with current foodservice and catering sales, it is always a smart idea to fine-tune already
successful operations by relying on customer and visitor research
conducted in local markets on an ongoing basis.

Professionally conducted market research and the data it uncov-
ers can shed a tremendous amount of light on ways to implement
a successful foodservice and catering program in locales where no
such hospitality services previously existed. Market research is an
essential first step if administrators want to expand currently inad-
equate restaurant and special-events venues—for example, by trans-
forming vending machines and dark basement cafeterias into the
bright contemporary cafés and catering venues that appeal to today's highly sophisticated foodservice customers and museum-
goers. Before costly renovation projects are planned, implemented,
or even considered, it's important for administrators to be able to
answer two questions. First, why is the format (or concept or menu)
being changed? And second, how will the changes make the institution's foodservice more responsive and better able to penetrate
targeted markets? Market research is one of the best ways to attain
appropriate insights and answers.

Market research allows cultural institution foodservice personnel
to gain candid, reliable, and valuable feedback from both current
customers and noncustomers, as well as to survey what the competition is doing in similar areas. Whether presented in the form of
specially developed questionnaires that elicit the likes and dislikes
of current members or as focus groups that allow a flow of candid
conversations between foodservice staff members and a selection of
customers likely to patronize an institution's dining operations, the
key findings of a market study can prove invaluable in the development of new restaurant and special-events programs that attract
strong sales, satisfy customers, and turn those customers and visitors into repeat users.

Market studies should be organized so that administrators can
discover the answers to the following questions, among others: To

what extent do current restaurant and special-events operations fulfill visitor expectations and requirements? Are current menu items those that visitors prefer to purchase and, if not, which offerings would they prefer instead? How do visitors rate the current style(s) of service? What is their price-value perception of foodservice operations? What type of service would visitors prefer if restaurant operations were changed or upgraded? Would customers prefer table service and, if so, what amount would visitors and members be willing to pay for such a service experience? How important to guests is the availability of catering services at the institution? What sort of catering options are best suited to current and potential guests? If we build a new additional café, what is the prospective demand and probability of success?

The following case studies describe how a variety of cultural institutions have conducted market surveys to evaluate the current state of their restaurant and special-events operations, as well as to explore the potential these operations have to improve in regard to both customer satisfaction and overall sales. Each cultural institution may have begun its market research process with a different agenda in mind, such as rating the effectiveness of current foodservice or catering offerings, or setting out to determine the feasibility of adding new foodservice amenities in a facility where none existed. However, the common thread that runs through these studies is the wealth of data obtained by the research process, enabling administrators either to improve hospitality services' current performance or to build a new or revised foodservice or special-events operations most likely to meet their market's particular preferences.

DETERMINING A CAFÉ'S PERFORMANCE

One such market research program was recently conducted for a botanic garden's visitor center. This popular institution commissioned a market survey with the express purpose of determining the strengths and weaknesses of an on-site café, the centerpiece of its foodservice operation and its strongest potential profit maker.

The market survey conducted here consisted of four steps. Phase one involved gathering data via three focus groups held at the garden—two with members and one with staff. The topics addressed by these focus groups included both staff's and members' perception of the current café foodservice program, as well as suggestions for future changes and improvement.

The second market research phase focused on the development of a questionnaire that was mailed to some 800 members of this cultural institution. A total of 245 surveys were sent back by mail, a better than 25 percent return rate.

The third step involved distributing a shortened version of the direct-mail questionnaire that was circulated among all café customers visiting the garden during a four-week period. A total of 94 completed surveys were gathered from these customers during this research phase.

The fourth and final research activity included a professional analysis of all completed surveys in order to provide garden administrators with viable options for planning the café's future foodservice offerings, as well as direction for any renovations or changes that would make it more responsive to its customer base.

This survey questionnaire had been specifically designed to give the botanic garden's administrators an accurate picture of how well the dining experience provided at their café fulfilled the expectations and requirements of its customers. Survey forms had asked respondents to evaluate their café dining experience according to 15 different criteria. These included freshness and tastiness of the food served, healthfulness of the food, the variety of menu offerings, food temperature, perceived price-value relationship, availability of children's menu items, cleanliness of the dining area, helpfulness of café staff, speed of service and table availability, ease of entertaining children in the café's atmosphere, the facility's ambiance, its location, and the speed of service through the cashier line.

Customer feedback generated by the survey gave the café's current operation an overall rating between "good" and "very good." The café received relatively high ratings for its convenient location, atmosphere, and cleanliness. Evaluations of food freshness, healthfulness, proper temperature, and taste were also relatively high, according to these respondents. As a result, market research was already helping administrators and foodservice managers learn exactly where they were hitting the mark in meeting customer expectations.

Functions of this café that affected families, however, received relatively low ratings from survey respondents, meaning that menu offerings for children and families' ability to manage dining experiences when children were in tow were areas that management needed to improve in order to increase customer satisfaction. Other problem areas that survey responses disclosed were the length of time it took to move through the café line and be checked out at the cashier station, with both being rated quite low. Determining ways to improve speed of service and checkout would, therefore, be among the first problems for management to tackle.

The garden's survey was also designed to gauge the importance to visitors of certain aspects of the café experience. Freshness and tastiness of the food and the facility's cleanliness were ranked highest in importance, along with the healthfulness of the products offered and seating availability. Respondents ranked services related to family issues, such as the availability of a children's menu, least important.

The "importance scores" system revealed that garden administrators needed to improve the perceived value of the café experience, as well as reduce the time required for customers to get through both the serving line and the cashier's station. Since the café scored poorly in terms of perceived value for money, another area of opportunity for improving guests' satisfaction was targeted.

RESPONDING TO FOCUS GROUP FEEDBACK

One-on-one feedback provided by focus group discussions also proved helpful in spotlighting certain strengths and weaknesses in the overall day-to-day operations of the café. The quality and freshness of baked goods such as scones and muffins merchandised at the café were not viewed as consistent from day to day. Customers revealed they had the perception that the café sold day-old baked goods—certainly not an impression likely to boost sales. The thickly sliced bread used in many of the café's sandwich offerings was also singled out as being difficult to handle, especially for children.

The use of disposable ware at the café was viewed by both members and visitors as ecologically unsound and inconsistent with the mission of the institution, as well as cheap-looking (a concern that was repeated by members and visitors alike in their survey responses). Another concern was the language barrier created by foodservice staff who were not fluent in English. The hectic, noisy atmosphere of the café during peak periods (particularly lunch), especially in comparison to the quiet air of respite that prevailed at other times of day, was also highlighted as a customer concern.

As a result of focus group findings, as well as responses to the written questionnaires, administrators at this botanic garden could now address the feasibility of making several relatively simple changes with its foodservice operator in order to improve customer satisfaction. These included training customer-service staff to be fluent in basic spoken English; maximizing the use of existing counter space to display available food items more attractively; forming a commitment to serve only freshly made scones, muffins, and other bakery items; and selecting alternative bread choices that were more thinly sliced and manageable, even in the hands of youngsters.

The menu mix offered at the garden's café had been a major area of investigation both on the questionnaire and during focus groups. As it could be assumed that the palate and menu preferences of the café's customers had changed in accord with current interests in ethnic specialties, bold flavors, and extensive variety, garden administrators needed to know members' and visitors' preferences regarding available entree items. Were any changes or upgrades necessary?

Market research discovered that (in descending order of popularity) chicken, hot sandwiches, wraps, pasta salads, and comfort foods rated highest among both members and visitors polled during recent research. Less popular, in order of preference, were "designer" pizzas, Asian foods, hot dogs, burgers, plain pizza, and Mexican items. Those surveyed further showed a marked preference for soups, healthful offerings, fresh fruit, baked goods such as muffins and bagels, and salads made to order. Rating low in popularity among the heath-conscious sample groups were foods such as high-calorie, high-fat french fries, potato chips, and desserts, including pies, cakes, cookies, and brownies. Coffees and teas proved to be the drinks of choice to complement menu offerings, according to those surveyed by this cultural institution, with bottled juices, soda, espresso, flavored coffees, and bottled waters being only moderately popular. Such feedback on menu preferences can provide solid evidence for cultural institution administrators as they plan the best ways to make future menu changes and facility concept alterations.

EVALUATING SERVICE OPTIONS

A third area of concentration for the garden's market research team was the type of service preferred by café customers. The scores for cafeteria style and food-court style (eat-in items) were highest, with the "grab-and-go" style (items packaged for takeaway) quite low on the preference scale. This allowed garden administrators to recognize that an important priority for future foodservice facility development would be to add an expanded food-court- or cafeteria-style operation.

The question of whether the garden café's management should offer table service was another option researched by foodservice administrators. Subsequent findings indicated that nearly half of the institution's members surveyed would be likely to patronize a restaurant that offered table service. Whether in fact administrators here would pursue the implementation of table service also depended in part on the optimal check average and price-value perception identified by the market study.

In the case of the research done for this cultural institution, members indicated a willingness to pay between $10 and $12 per capita at a table-service restaurant, compared with approximately $7 per meal at a food court or café concept. The issue of price-value perceptions proved to be the most sensitive addressed by this market research process. Overall, garden members and visitors saw café prices as too high for the value received. Although prices at the café were within the range of those set by other cultural institutions in the vicinity, the surveys found that other factors influenced its customers' price-value perception. These included the fact that meal items were then being presented on plastic disposable servingware with disposable utensils in a cafeteria ambiance. The design and attractiveness of serving counters and display areas were also rated unfavorably in comparison to other cultural institutions' foodservices in the same metropolitan area, as well as commercial operations offering similar fare.

Market research findings further supported the position that this café should not lower its meal prices, since current foodservice demand already exceeded capacity. In addition, garden administrators learned they would do well to avoid any further upscaling of the menu offerings, because the price increases necessary to support the introduction of higher-priced meal choices would not be well received by the institution's members or visitors.

LEARNING TO READ RESEARCH SNAPSHOTS

The market research survey undertaken to evaluate the strengths and weaknesses of this garden's café operation also provided a variety of interesting "snapshots" that indicated what should stay, what should go, and what should be revised for improved market penetration and customer satisfaction.

Suggested changes to enhance overall operations included improving signage in general and developing a less confusing ordering process, with prices more prominently displayed and laminated color photos of each food item placed above order areas. Other necessary changes included a more organized approach to cashier station operations in order to eliminate long lines. Findings also indicated that an extension of café hours should definitely be considered, with the facility opening earlier in the morning and closing later in the evening, particularly during summer months, when visitor traffic was at its peak.

A quicker table turnaround time was also suggested, with tabletops cleared and floors swept in a more efficient and timely manner so that customers would no longer have to stand around, trays in hand, waiting for a bussed table to open. Survey respondents additionally indicated that the indoor dining area would benefit from enlargement; the outdoor patio would be more comfortable if it was screened to eliminate bugs and if umbrellas were placed over outdoor tables to provide cooling shade. A switch to permanent dishes and utensils was recommended as soon as administrators could determine a suitable area to install dishwashing equipment.

Suggested menu improvements that would not require an increase in meal prices included a variety of potentially workable ideas. These included a free-refill policy on beverages, more use of fruit and vegetables already being grown on the garden's own grounds, a greater number of "heart-healthy" options overall, as well as more healthful options for children, allowing ingredient substitutions in items such as salads and sandwiches to accommodate some customers' dietary restrictions, expanding breakfast selections by including an omelet station, and offering a greater number of foods perceived to be fresh and unusual.

Other viable options for expanding and enhancing foodservice operations at this garden included the addition of a table-service section in the main dining area, expansion of the seating capacity, providing chairs that were more comfortable, soft classical background music, a fountain and sculpture to improve the ambiance, introducing a food cart to let customers purchase coffee and tea as well as to-go desserts without having to stand in long lines, and a switch to automatic doors at the entrance to the outside patio dining area for the comfort of patrons whose hands were occupied carrying trays.

Researching to Guide Growth

A second case study shows how market research helped administrators at a wildlife sanctuary make the right decisions about how to go about the process of expanding their current foodservice operations, which were then offering visitors very limited on-premises dining options. The initial challenge was to determine the extent of food and beverage offerings that would be acceptable, given the fact that this sanctuary's visitors had evidenced both an overall desire to purchase a selection of midscale foods and beverages and a sensitivity to the environmental impact that the presence of these products, along with their containers, packaging, and serving equipment, would have on the total park experience.

Prior to conducting its market study, the sanctuary's foodservice program offered a small selection of food and beverage products, including canned carbonated soft drinks, bottled water, and packaged juices, primarily distributed from on-site vending machines, along with candy and popcorn that were available at the visitor center.

Subsequently, research revealed that foodservice's capture rate of the available daily market was only 36 percent, which was surprising, given a reported annual attendance of 100,000 guests and the remote location of this institution. Visitors' average daily per capita spending was just $0.40.

By making comparisons with cultural institutions with attractions and audience demographics similar to those of the sanctuary, where per capita food and beverage expenditures were in the $2.00 range, administrators at this cultural institution knew that there was tremendous room for improvement, possibly by as much as 100 percent, based on an expansion of food and beverage services that was appropriate to available visitor volume and disposable income. The remote location of the sanctuary, combined with the long time visitors tended to stay and their tendency not to have eaten prior to arrival, all provided a good opportunity for this institution to more than double its foodservice sales from an annual total of less than $40,000 to over $100,000.

The market research team here first set out to identify guest and volunteer dining patterns and desire for selected types and styles of foodservices. This information would allow management to ascertain visitors' inclinations to patronize different types of on-premises foodservice operations and their specific preferences for food products not then being offered. Research information would also help sanctuary administrators determine the financial viability of implementing an expanded but still limited on-premises foodservice program that would effectively meet members' expectations.

The main instrument of research in this study was a survey questionnaire, which was distributed to randomly selected visitors and volunteers during a one-week period during midsummer. Each participant received a complimentary cold beverage as an incentive to complete the survey, as well as a verbal thank-you for participating in the market research process.

Statistics were ultimately drawn from responses to 120 guest surveys and 6 volunteer surveys; respondents answered 17 questions on each survey. Regarding frequency of visitation to this sanctuary, more than 70 percent of the respondents reported that they were making their first visit, with only 7 percent reporting themselves frequent visitors (visiting the wildlife grounds once a month or more). It is interesting to note that this group of frequent visitors tended to bring their own meals from home, further highlighting the potential market for an enhanced on-site foodservice program. Asked about the possibility of future visitation, a very high number (93.4 percent) of guests queried said they planned to return to the sanctuary in the near future.

Question three of this survey addressed the issue of dining prior to arriving at the park. Seventy-six percent of visitors did not eat prior to their visit, and over 79 percent of all respondents reported visiting the sanctuary during the traditional lunch period (11 A.M. to 2 P.M.), providing additional evidence that an audience definitely existed for an expanded foodservice program.

DISCERNING DINING PATTERNS

The next two survey questions linked the issue of prior dining to visitation patterns and elicited the fact that more than half the visitors who stopped to eat prior to their arrival at the sanctuary had planned this stop, although one-third of this group responded that having a full foodservice program available at the institution would have influenced their decision whether to dine prior to visiting.

More than half (56 percent) of visitors also noted that they intended to stop to eat immediately after leaving the sanctuary, with a high proportion of those visiting the park in the morning (61.7 percent) and those visiting over the lunch period (59.7 percent) planning to dine immediately after departing. Answers to question six also demonstrated that the longer a visitor stayed at the park, the more apt that guest was to eat out immediately after leaving the grounds, another indication that a high proportion of park visitors would be amenable to patronizing a foodservice operation on-site.

The next survey question revealed that more visitors in the morning (52 percent) and during the lunch period (50 percent) would be willing to change their eating plans if foodservice were available on-site than those visiting the park in the afternoon (33 percent). Overall, fewer women's dining patterns were likely to be influenced by the availability of food, and 54 percent of visitors' dining patterns would not be significantly influenced if food was available at the park. By contrast, two-thirds of park volunteers responding to the survey indicated that the availability of foodservice would influence their dining patterns.

The same research study further revealed that more than three-fourths (78 percent) of visitors reported having had something to eat or drink during their visit. Looked at according to gender, this response broke down to 83 percent of women and 69 percent of men reporting the purchase of foods or beverages while at the sanctuary. Highest food or drink consumption was (as could be anticipated) during the lunch period, when 85 percent of visitors either ate or drank something, with transactions dropping to 75 percent of guests during the morning and 67 percent during the afternoon hours.

The majority (67 percent) of visitors who either ate or drank during their visit to the sanctuary brought the item they consumed with them from home. Only 27 percent of the visitors who ate or drank during their visit claimed to have purchased the food or beverage from the sanctuary's on-premises vending machines or store. Almost 14 percent of those eating or drinking during their visit admitted to buying their food and/or beverage item on the way to the sanctuary and carrying purchases in with them.

The majority of those eating or drinking during the morning here tended to bring foods or beverages from home. About 75 percent of those eating or drinking during the traditional lunch period had brought snacks from home, but only 38 percent of those eating or drinking in the afternoon had brought foods or beverages from home. The survey concluded that, overall, guests visiting the park for more than two hours were the most likely to bring foods and beverages from home, and that this group was the most likely target audience for on-site dining services.

Responses to another question on the survey identified the buying habits of visitors who were purchasing what the sanctuary did have to offer by way of limited food and drink options—canned carbonated beverages, bottled waters, and packaged juices available from three on-site vending machines, and candy bars and popcorn sold in the visitor center.

The overwhelming majority of items purchased on-site (98 percent) were obtained from vending machines, and 25 percent of visitors bringing food from home supplemented these items with beverage purchases from vending machines. Bottled water ranked as the most popular beverage purchase. Volunteer staff reported no purchases of foods or beverages from the visitor center.

ELICITING INPUT ON PROGRAM PREFERENCES

A number of survey questions were designed to obtain feedback regarding visitor and volunteer preferences for an improved foodservice program. A substantial number of respondents (56 percent) indicated a preference for a program of light-meal offerings, such as cold deli sandwiches. A beverages-only menu was the next most popular choice, followed by hot sandwiches and snack foods. It should be noted that the option of having no food and beverage selections placed last. As might be expected, men expressed a desire for hot foods, such as hot dogs, more often than did female respondents. When all responses had been analyzed, it was determined that light-meal offerings were the type of foodservice most preferred by both the visitors and the volunteers polled for their input.

In conjunction with the question of what style of foodservice to feature, survey respondents were also asked about their preference for a particular style of service. The majority of those polled (53 percent) indicated a preference for counter service with limited seating. Other options receiving less support were (in descending order) a convenience store with limited seating, vending machine service only, no foodservice, and a convenience store offering no seating. As can be surmised from these responses, interior seating was strongly preferred over outdoor dining, and women voiced a higher preference for interior seating than did men.

Another question asked respondents to indicate the specific foods they would most likely purchase should these items be made available at the sanctuary. The desired menu mix as outlined by respondent references included cold beverages, bottled waters, and juices at the top of the list. Second-tier choices included ice cream and yogurt, fresh fruit, and packaged, premade deli-style sandwiches. Other food offerings that visitors were most likely to purchase included packaged fresh salads, potato chips and popcorn, submarine sandwiches, and wrap-style sandwiches.

Those surveyed noted that they preferred different food items depending upon the time of day (with cold beverages ranked as the most popular menu choices at all times). Deli sandwiches, for instance, were favored both in the morning and in the afternoon, while ice cream and yogurt were preferred during afternoons only.

Despite the limited foodservice offerings at the sanctuary, 61 percent of all respondents rated their experience there as "above expectations," with no one rating his or her experience as "below expectations." One-quarter of the queried population felt their experience at the park would be enhanced by the availability of food, with those who spent more than three hours at the sanctuary most inclined to feel that their park experience would have been more enjoyable had food been available.

All volunteers responding were supportive of additional food products being offered at the sanctuary. But their reaction was decidedly mixed about having food and/or beverages available for purchase at all (keep in mind the ecological consciousness of those volunteering to work at a wildlife sanctuary), with 16 percent opposed to merchandising any foods at all.

Researching visitor tendencies in relationship to the potential enhanced foodservice program allowed sanctuary administrators to obtain a clear picture of how to revamp their on-site program in order to meet visitor and volunteer preferences and expectations. Bottled waters, cold sodas, and juices were identified as the preferred foodservice items, with light-meal options such as cold sandwiches served from counters determined to be potentially the most profitable additional menu options.

Matching Development to Visitors' Perceptions

Our third market research study, conducted for a large aquarium, demonstrated how this sort of analysis can serve to strengthen an existing foodservice program, as well as provide direction for increasing catering and special-events opportunities.

A five-page survey administered on-site to visitors to the facility, as well as members of the public at an adjacent shopping mall, tackled the question of the extent to which the aquarium's limited foodservice program was penetrating its existing market. Additionally, administrators here wanted to know if the cost of developing new foodservice and catering opportunities could be justified. By understanding just where their hospitality programs stood now with guests, these administrators would be better equipped to modify their foodservice marketing efforts and their approach to gaining greater market share.

The first segment of the research survey created for this institution set out to create a profile of the facility's current customers and their buying habits, dining customs, and preferences. Initial research indicated that a high percentage (75 percent) of visitors planned on returning to this aquarium again in the future; more than half reported that they had not stopped to eat on their way to the institution, and of those who did stop to dine prior to visiting, 42 percent said they would have postponed eating until arriving if they had known foodservice was available on-site.

However, only one in five current and previous visitors said they had purchased either food or drinks while at the aquarium, with the majority of these purchasing only a beverage. What's more, during a typical day, more than half (59 percent) of the aquarium's visitors were on-premises during lunchtime (between 11 A.M. and 2 P.M.). This discovery indicated a major weakness in foodservice's marketing approach, since fully 30 percent of visitors who came to the aquarium at lunchtime were not consuming *any* foods or beverages while on the premises and, of the visitors who did make a foodservice purchase, 78 percent were buying just water or soda. This profile of current customers pointed toward some exceptional opportunities for the aquarium's foodservice to offer a relatively wide variety of foods and beverages to visitors, particularly during the peak lunch period.

The next segment of survey research dealt with customers' food and beverage preferences. The majority of respondents indicated very strong preferences for light, café-style foods, such as salads and cold deli-style sandwiches, with fast foods and casual-dining menu items such as ribs and gourmet pizzas noted as second-tier preferences. Joined to this list of preferences was a stated desire for a casual-dining program, including counter service, to be available from foodservice, with meals taking anywhere from 45 to 60 minutes to complete. This aquarium's respondents more specifically indicated that a counter-service café was preferred for lunch, while a casual-dining restaurant was the top choice for evening meals. Notably, interest among respondents for purchasing alcoholic beverages was very low.

INDUCING CUSTOMERS TO DESCRIBE THEIR IDEAL FOODSERVICE

Asked further to construct an "ideal" menu for the aquarium's foodservice, respondents showed their strongest preference for made-to-order deli-style sandwiches. This product line clearly outdistanced anything mentioned on the second tier of food choices, including made-to-order salads, pizza, and fruit plates. This institution's survey group also showed a strong preference for cold beverages, including bottled waters and juices, and snacks such as ice cream and frozen yogurt as complements to their main selections. Surveyed customers further noted that they preferred fresh, made-to-order meal choices rather than premade and/or packaged products. In fact, in this instance, only packaged hot dogs received any support from surveyed respondents.

When questioned about pricing, always a sensitive issue, the survey revealed that respondent's average lunch and dinner expenditures were $5.29 and $10.53, respectively, indicating aquarium visitors' ability and willingness to pay for foodservice products that provided an attractive price-value relationship. The demographic makeup of the surveyed visitor group—with an average age of 41, an average income of over $36,000, the majority married, educated through at least some college, and having children under 19 living at home—also supported the assumption that this customer base could support foodservice per capita expectations in the range of $3.00 per visitor, a full dollar above average sales at most comparable cultural institutions.

EXPANDING MARKET PENETRATION

This aquarium's easily accessible location, its visitors' average stay of 1.67 hours, and guests' tendency not to have eaten prior to arriving all indicated a promising opportunity for this institution to generate substantial foodservice sales by taking the appropriate measures to expand its limited program offerings.

In short, the market survey revealed that the aquarium's visitor base consisted of an interested clientele with significant disposable income, who were amenable to patronizing an on-site restaurant for lunch and/or dinner at some point during their visit. Other expectations included friendly service staff and high-quality food and service, preferably presented in a relaxing café during lunch and a casual-dining restaurant during dinner. Men, local residents, and those visiting with children were the primary customer groups who felt their experience at the aquarium would have been even more positive had such a foodservice program been available.

The solution to greater market penetration for this cultural institution's foodservice was to expand existing restaurant offerings by adding a variety of new items, including made-to-order deli-style sandwiches and cold beverages. This main point of sale would best be supported by other flexible food and beverage stations, such as mobile carts.

Assessing Catering Demand

The second part of the research survey conducted for the aquarium attempted to answer the question of whether the facility could support and sustain catered events booked by local groups and individuals. Questions asked in a survey of local residents and visitors shaped a profile of what the local catering market and its potential looked like.

Almost half (47 percent) of respondents said that they catered events outside their homes, a small percentage of these on a monthly or more frequent basis. The average number of people at these catered events was 24, with the majority (86 percent) of events consisting of 40 people or less. Tourists indicated that they were least likely to hold a catered event at the aquarium, though 47 percent of the local market expressed doubt that they would call upon the cultural institution's catering program for special events. Some respondents expressed the opinion that its location "was too far away" and "not convenient enough" to make it a preferred site for a catered affair.

However, there did seem to be adequate support among local residents for the aquarium to consider proceeding with the development and marketing of catering services. While not as enthusiastic in their response to catering as they were to the expansion of restaurant offerings, the surveyed group indicated that it was at least open to booking catered events at the cultural institution. Information uncovered via market research gave a strong rationale to exploring the possibilities inherent in developing a catering/special-events program.

The next challenge to be tackled by aquarium administrators (though not addressed by this market survey) was the further identification of the style of catered events to which local residents would be receptive. A second, catering-focused market research study would be the next logical step in the investigative process leading toward the development of a successful catering/special-events program at this cultural institution.

CHAPTER 10

UBIT: A Food and Facilities Tax Primer

Jeffrey M. Hurwit*

Most cultural institution administrators probably have some familiarity with the basic premise of the UBIT (Unrelated Business Income Taxation) regulations. Simply put, if a museum makes money on any activity (such as foodservice) unrelated to its tax-exempt purposes, the profit from that activity is taxed at for-profit business rates. More technically, unrelated business income is net income derived from a trade or business regularly carried on that does not contribute importantly to the accomplishment of a museum's tax-exempt functions.

This chapter focuses on the three major exclusions to UBIT relating to cultural institutions' foodservices. These exclusions likely mean that most food and facilities profits earned at cultural institutions are not taxable. However, UBIT rules are continually being

*© 1999 New England Museum Association. Originally printed in *NEMANews*, summer 1999, volume 22, number 4. Jeffrey M. Hurwit is an attorney in Boston, MA, whose tax firm, Hurwit & Associates, counsels arts, educational and tax-exempt organizations throughout the United States and abroad.

refined, so institutions should seek expert advice from a qualified professional.

VENDING MACHINES TO FOUR-STAR MEALS: NONTAXABLE FOODSERVICES AS A CONVENIENCE TO VISITORS AND EMPLOYEES

Whether administrators oversee a snack bar, café, or fine-dining restaurant, whether foodservice staff sell candy bars or crème brûlée, on-premises food sales to cultural institution employees and visitors contribute to accomplishing tax-exempt purposes for two reasons: They allow visitors to devote more time to a museum's educational exhibits, and they enhance efficient museum operation by enabling staff to remain on-site throughout the day. Food sales, like water coolers, restrooms, and exhibit-room benches, facilitate and enhance the museum experience. Thus, resulting revenue is not taxable (Revenue Ruling 74-399 [1974]).

While this may be the current legal conclusion under most circumstances, small facts can make big legal differences. For example, if a dining facility is accessible not only through a cultural institution but also through a door opening directly from a street, then such facilities have been held by the IRS to be primarily not for visitor convenience, but for general public use and, therefore, taxable (ibid.). Thus, museum architectural designs and facility placement and access plans may be affected by UBIT regulations. A recent case has now imposed additional limits on this "convenience exclusion." As described in IRS Technical Advice Memorandum 97-20-002 (1996), a museum opened an upscale restaurant larger than needed for visitors and staff. The IRS determined that since this facility was designed partly as a public restaurant and was advertised regularly in magazines, the convenience exclusion did not apply. Administrators should note that the fact that this restaurant's patrons did not have to pay museum admittance fees was one factor that influenced the IRS decision, but was not itself determinative.

To museums' benefit, on the other hand, the rule of "fragmentation" applies when administrators calculate net foodservice income. Cultural institutions may "fragment" restaurant sales (separately record sales to museum visitors, staff, and the general public) and pay taxes only on sales to the general public.

USE OF MUSEUM FACILITIES BY OUTSIDE GROUPS FOR EDUCATIONAL PURPOSES

Increasingly, cultural institutions are using their facilities to host outside business and social affairs. The legal issue is then whether an event is held primarily for business purposes or for educational purposes (in which case food and entertainment are considered to be incidental).

Suppose an outside sponsor asks a museum to create an educational program for its participants, focusing on an exhibit, lecture, or tour, and incidentally asks that food and other services be provided. Because such an event contributes to accomplishing the museum's purposes, it is not subject to tax (IRS Technical Advice Memorandum 97-02-003 [1996]).

However, if an event is primarily focused around, for example, a cocktail reception, dinner-dance, business meeting, or awards ceremony, then the educational aspects are considered to be secondary to other business purposes and therefore are not "substantially related" (*Madden v. Commissioner,* T.C. Memorandum 1997-395). That interpretation may prevail even if an exhibit opening, tour, or other educational component is included in such a event.

PASSIVE RENTAL OF MUSEUM FACILITIES

The rental of real property, including special-events facilities, is not considered the "active conduct of a trade or business" by a cultural institution but rather the "passive" receipt of revenue. So-called passive revenue has long been excluded from UBIT (Internal Revenue Code Section 512(b)(3)). Thus, even if an exclusion does not apply due to an event's lack of educational content (as in our previous example), a cultural institution will likely still avoid UBIT if it simply rents a function room or property without providing additional services. However, providing additional services, such as labor, food, catering, or linens, is considered actively conducting a business, and the rental arrangement will likely be subject to tax (IRS Technical Advice Memorandum 97-02003).

As is evident, the factual variations and legal distinctions relating even to these three exclusions leave many open questions, but we hope that this brief primer gives cultural institution administrators a sense of the legal parameters for the majority of UBIT situations they will encounter.

Food Safety: Too Often Ignored Until Too Late

Every cultural institution that operates a café, restaurant, snack bar, food cart, or kiosk or that stages special events with catered food and/or beverages needs to be aware of and concerned about food safety. This chapter will address some of the proactive steps cultural institution administrators can take to reasonably ensure that visitors, guests, and staff remain safe from food-borne illness and that their institutions are protected against the liability issues and negative publicity that often result from lapses in food safety.

HACCP: WHAT IS IT?

HACCP, or Hazard Analysis of Critical Control Points, has as its fundamental purpose to help foodservice operators identify potential problem areas in their facilities and create a system of procedures and solutions to avoid unsafe practices.

There has been tremendous growth in the use of HACCP principles, which stress time and temperature controls. Time and temperature abuse takes place anytime food has been allowed to sit for too long at temperatures that promote the growth of microorganisms. Contributors to food-borne illness include:

- Failure to hold or store food at required temperatures.
- Failure to cook or reheat foods to temperatures that will kill microorganisms.
- Failure to cool foods properly.
- Preparing foods a day or more before they are to be served.

Another important cause of food-borne illnesses is cross-contamination between foods, facilitated by poor food-handling procedures and poor sanitation in regard to food-contact surfaces. Among the key concerns in this area are kitchen staff (food handlers), who most often cause cross-contamination by:

- Not washing their hands frequently when working with food.
- Touching raw foods and then prepared items without washing their hands in between.
- Allowing raw or contaminated food items to touch or drip fluids onto cooked or ready-to-eat items.

WHAT SHOULD CULTURAL INSTITUTION ADMINISTRATORS DO?

During operations meetings with a foodservice operator (or manager), administrators should ask to see copies of the last several health inspection reports from the local city or county health department. (The frequency of these inspections varies from annually to several times each year, though they are usually not conducted on a regular basis.)

Based on our experience, we recommend that cultural institutions consider engaging the services of an independent food safety inspection service. There are many such firms throughout the United States that are hired by restaurant and hotel chains and independent restaurant and foodservice operators to inspect their facilities because these businesses do not have the in-house resources to do so and/or they do not want to rely on inspections by a local health department.

Regardless of how well an administrator thinks a foodservice manager or operator is handling food safety and sanitation issues, a proactive institution that wants to be certain that its in-house foodservice operations are always operating at the safest possible level will engage an independent inspection service. If foodservices are outsourced, the operator will often share the cost of this service (about $300 to $500 per inspection, depending on the size of the foodservice facilities, with inspections usually conducted quarterly). Occasionally, an operator will agree to pay the entire cost of a single inspection if the rating is below an agreed-upon baseline.

It is also vital for administrators to ask their operators or managers if they have completed a food safety training course (ask to see a certificate or similar document); if the individual has not, we recommend that he or she do so as soon as possible. (Some states and/or local governmental agencies require such training.)

In addition, administrators should ask about food safety training for all food handlers working in the institution's foodservice operations. It is well worth the investment to support (financially, if necessary) whatever training and materials are required to ensure that foodservice staff are following accepted industry standards in this regard.

HOW CAN ADMINISTRATORS STAY ABREAST OF THIS ISSUE?

The best way for administrators to receive all the information they need about food safety, HACCP, and related matters is for their institution to join their state's restaurant association. If on-site foodservice is provided for visitors, guests, and/or staff, then administrators should look at this membership the same way as they do enrollment in AAM, AZA, ASTC, MSA, or other state or regional museum trade associations.

Restaurant associations all provide newsletters, reports, and printed materials covering food safety and related issues. In addition, administrators can use their restaurant association to access a list of foodservice industry trade publications to which they can subscribe, as well as obtain listings of independent food safety inspection firms in your state or local area. This is a very small investment with a very large potential return.

WHAT ABOUT FOOD SAFETY AT CATERED EVENTS?

If an institution exclusively uses in-house management and staff to provide meals at catered events, then the above comments and recommendations apply. If an institution uses a list of approved caterers and/or is open to any outside caterer, then administrators should be aware of the following issues and concerns:

- It is not advisable to have an open catering policy.
- All cultural institutions should have a list of approved caterers (regardless of how many caterers are on the list) that can be updated annually based on the caterers' ability to meet minimum standards, including those pertaining to food safety.
- There should be a formal contract that includes food safety standards between any approved caterer and an institution.
- Administrators should also receive a copy of the caterer's certificate of insurance, to be sure the institution is protected from any and all liability arising from the actions of the caterer and, possibly, the caterer's subcontractors and/or suppliers.
- Administrators should ask their caterer for a copy of its health permit or license covering its central kitchen (and vehicles, if applicable). Administrators should be sure they get an updated copy at each renewal period.
- Before approving a caterer, administrators should ask to see copies of health inspection reports of its central kitchen for the most recent 24 months. Ask for an explanation of any

deficiencies. (Some health departments have web sites where this information can be accessed.)

■ Administrators should consider personally touring and inspecting any approved caterer's kitchen to satisfy themselves that it meets the same high standards as the foodservice facilities at their own institution.

■ At any cultural institution where food is prepared or served, food safety and sanitation need to receive the same priority and attention to detail as the upkeep of the institution's collections.

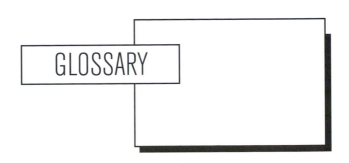

GLOSSARY

Administrative and General Expenses

A foodservice operator will assess each of its client foodservice operations an administrative and general expense fee or charge usually equal to a percentage of gross operating revenue, ranging from 3 to 5 percent depending on the size (dollar volume) of overall foodservice operations (the higher the gross operating revenue, the lower the percentage). These are also known as "general and administrative expenses" or "G&A."

Agreement

The term is used interchangeably with the word *contract* in the on-site foodservice industry. An agreement clearly sets forth all the business terms and conditions covering the relationship between a client and a foodservice operator, including but not necessarily limited to recitals that provide an overview of the client institution, the operator, and the services contemplated under the agreement; term and termination; financial, capital investment, and payment terms and conditions; services; what is provided by the client and what is provided by the operator; insurance; indemnity; security; health and safety; legal terms and conditions and exhibits covering the space allotted for foodservice; menus, prices, catering terms and conditions; financial reporting format; sample customer survey forms; minimum food specifications; and related factors.

Alcoholic Beverage Cost

The ratio of dollars of alcoholic beverages purchased to revenue, expressed as a percentage.

À La Carte

Menu items that are priced individually.

Approved Caterer

A local caterer that has met certain standards, including having suitable experience, possessing a health-department-approved catering kitchen and vehicles, carrying appropriate insurance as evidenced by a certificate of insurance provided to the cultural institution, and signing an approved caterer agreement with the institution to provide food and beverage catering services on-site. Only approved caterers (and the in-house caterer if one exists) should be able to provide catering services at the institution.

Approved Caterer Agreement

The agreement (contract) that is signed by an approved caterer, usually a year-to-year agreement that may be canceled on 30 days' notice and which details the terms and conditions under which the off-premises caterer will provide its services at an institution. This agreement usually provides for a commission payment to the institution on food and beverage (only) sales, and sometimes includes a minimum annual dollar guarantee to the institution, as well as the guarantee of an annual dollar donation of services (i.e., *pro bono* services valued at an agreed-upon amount).

Approved Caterer List

Usually a short list of local caterers that meet an institution's minimum standards. This list will only include suppliers who have executed an approved caterer agreement and provided a certificate of insurance to the institution.

Back of the House

The food storage, office, staff, production, and preparation areas (kitchen areas), as opposed to front of the house, which includes the serving and dining or public areas.

Base Fee

The minimum dollar amount received by a foodservice operator under a management-fee type of agreement.

Branded Concepts

National, regional, and/or local trademarked or well-known concepts that can be incorporated into a foodservice program via a licensing or franchise arrangement. Examples include Pizza Hut, California Pizza Kitchen, Burger King, and Taco Bell. Not all branded concepts will license or franchise their foodservice program. Some branded concepts can be operated in cultural institutions but have to be run by an approved franchisee of the franchisor. McDonald's and Domino's Pizza are examples.

Branded Products and Costs

The products that must be used and/or the cost of supplies and products that must be purchased from a specific vendor as part of a franchise or licensing agreement. Branded products can also include everyday food and beverages sold in a restaurant/café, including but not necessarily limited to soft drinks, bottled and canned beverages, condiments (ketchup, mustard, etc.), and snack foods such as chips and cookies. These branded products are readily available from wholesale distributors that service the restaurant and foodservice industry.

Budgeting

The process of forecasting the expenses and revenue of foodservice operations over future months and/or years.

By-the-Ounce

A program whereby customers make their own salads or sandwiches or choose servings of other selected menu items, such as soup, and are charged by the ounce, so that those who eat lightly pay less than those who take hearty portions.

Café

Normally considered a small restaurant, casual in atmosphere (and decor), offering table service and/or self-service food and beverages.

Cafeteria

A self-service restaurant where customers move through a straight-line serving counter and/or go to individual food and beverage serving stations within a serving area (this latter design is called a scramble or scatter system) and either take prepared and plated foods or beverages or are served by a foodservice staff person (line server).

Call Brand

An alcoholic beverage product called for by a customer using its brand name.

Cannibalization

In a multi-operation environment, the process by which new menu items and/or concepts eat away at the market share of existing products/concepts.

Capital Investment

The sums of money invested for improvements in an existing or planned foodservice facility in a cultural institution. As a general rule of thumb, an operator will invest $100,000 per $25,000 of projected profit from foodservice operations at cultural institutions. Traditionally, operators will invest more dollars at a lower ratio in cultural institutions with several million dollars of annual revenues and where such investment is likely to materially (and positively) impact revenues and net profit.

Capture Rate

The percentage of total visitors who dine at an institution's foodservice operation(s). Ideally based on meal purchases only, rather than total transactions that include snack and/or beverage purchases.

Cash Over or Short

All foodservice operators balance the cash collected at their points of sale against cash register tapes. A day's reading is determined by taking the beginning register reading at the start of the day or a cashier's shift and subtracting the ending reading. If there is more cash than indicated by the total on the tape, then there is a "cash over." If there is less cash than the amount indicated on the tape, then there is a "cash shortage." Cash overages and shortages, as a rule of thumb, should not exceed 0.1 to 0.3 percent of total gross cash revenues.

Cashier

The foodservice staff person who operates a cash register and handles cash transactions with foodservice customers.

Cash Box

On occasion, a cash box (a metal or plastic box that has compartments for change and several denominations of bills) is used to hold receipts from sales of foods and/or beverages at certain high-volume foodservice operations where power might not be available to operate an electronic cash register. While cash box use is discouraged, the cash received can be audited and accounted for if the products sold are under strict inventory control. If, for example, a cash box is used at a cart selling hot pretzels, if the pretzels are inventoried at the beginning and end of the seller's shift, the consumption will be known and the number consumed times the selling price should equal the cash received. The seller should also note any waste or spoilage to ensure accurate cash box accounting.

Catering

The provision of pre-arranged food and/or beverage services that meet the specific needs of any group or organization, including your own institution.

Central Kitchen

A single production area, on-site or off-site, where foods are produced and from which they are transported to one or more preparation and/or service areas. Also known as "central commissary."

Check Average

Total restaurant revenue divided by the total number of daily customers for separate day-parts (breakfast, pre-lunch period, lunch period, afternoon period, and/or dinner period).

Client Liaison

The client or owner representative who has administrative responsibility for foodservice operations at a location. This person interacts with the operator if foodservice is outsourced and represents the owner's employees, staff, and visitors in dealing with the operator. This person may also have contractual oversight responsibility for negotiating and amending a contract between the institution and the foodservice operator.

Client/Owner

The proprietor of a cultural institution, school, college, public attraction, office building, or other location where a foodservice operation(s) is located.

Coffee Cart

Portable or semi-stationary self-contained serving cart offering coffee, espresso, and related coffee beverages, with the possible addition of cold bottled beverages, baked goods, and other snacks.

Collective Bargaining

The process that establishes conditions and wages acceptable to both union members and management. Hourly foodservice employees at some cultural institutions are covered under collective bargaining agreements.

Commercial Foodservice

Operations such as restaurants that compete for customers in the open market. A restaurant at a museum that has direct street access (can be accessed without going into the institution or without admission) could be considered a commercial foodservice under this definition.

Commission

The percentage of gross operating revenue paid to the client or owner by the foodservice operator for the right to provide foodservices in their venue. It can be a set percentage or sliding scale, increasing as gross revenue increases.

Concept

An element of a foodservice operation that contributes to its function as a complete and organized system serving guests' food needs and expectations. This includes menu, decor, merchandising, signage, staff uniforms, service style, equipment and supplies, and food displays.

Contract

Used interchangeably with the word *agreement* in the on-site foodservice industry.

Contract Foodservice Operator

A foodservice company that specializes in operating cafés, cafeterias, food courts, and restaurants (and catering and related food and beverage services) in cultural institutions, public attractions, corporate offices, schools or universities, convention centers, stadiums, hospitals, or similar facilities where the owner outsources this service. This provider may exclusively operate only these types of foodservice facilities, or may also be a commercial restaurant operator, hotel, or off-premises catering company that also is a contract foodservice operator.

Contract Types

When outsourcing, most cultural institutions will contract under either a profit-and-loss or management-fee type of agreement.

Controllable Expenses

One or more of all payroll, direct operating, advertising/promotion, utilities, administrative and general, repair and maintenance, and similar expenses that are under the control of a foodservice operator.

Cost of Sales

The dollar total of all food and beverages actually used in a particular month or accounting period, based on the beginning inventory and adding purchases, minus ending inventory. Also called "cost of goods sold."

Cover

A customer in a dining facility. Cover counts are synonymous with customer counts.

C-Stores

Cash-and-carry convenience stores that sell a variety of food, beverages, sundries, and groceries.

Cuisine

Food cooked and served according to the traditions of cultures from around the world.

Cultural Institutions

Museums, aquariums, zoos, botanic gardens, and historic mansions. All these public attractions are usually operated by a not-for-profit entity (with possible ownership by a local municipality).

Customer

The visitor, staff person, guest, or other individual who patronizes a restaurant, café, cart or kiosk, and/or catering service.

Customer Counts

The number of cash register (point of sale) transactions. Note that the number of cash register transactions does not provide the exact number of "customers" but the number of transactions, which is divided into the total revenue to determine the check average (or average check). It is more accurate and preferred if point-of-sale equipment can track the actual number of customers served.

Day-Part

A portion of a day when a foodservice serves breakfast, lunch, or dinner, as well as the time periods between breakfast and lunch and between lunch and dinner.

Demographics

Statistical profile of the characteristics of a cultural institution's visitors and/or staff.

Direct Operating Expenses

Those expenses directly attributable to food preparation and customer service, such as uniforms, laundry, linen rentals, linen replacement, china, glassware, flatware, and similar items and services.

Discounts (Average Amounts)

Commonly, employees (staff) of a cultural institution and volunteers receive a discount of at least 10 percent (but sometimes as much as 30 percent) off posted menu prices. Occasionally, members also receive 10 percent discounts. Discount sales should be tracked by administrators in order to determine the true cost to the institution.

Electronic Cash Register

Records the menu items sold at individual selling prices, as well as other management information such as customer counts and server identification (if table service is provided). Same as *point-of-sale (POS) system.*

Employee Meals

These are meals consumed by the foodservice staff. Depending on local, state, and federal tax rules and regulations currently in effect, these meals may have tax consequences for foodservice workers, so employee meals are not necessarily the same menu items offered to the public by a restaurant and can be selected by an operator because, though nutritious, they have a reasonable food cost.

Exclusive Rights Agreements

Cultural institutions may have exclusive pouring rights agreements with a soft-drink manufacturer (or other vendor or manufacturer of a food or beverage product), whereby the institution must exclusively use this manufacturer's branded products at specific contractual prices. These agreements typically provide marketing funds, as well as cooperative advertising and promotion for the manufacturer and the institution.

External Catering

Catering services provided to groups and organizations (such as for-profit and not-for-profit corporations) and others that, while potentially related or associated with a cultural institution, are not part of it.

FF&E

Furniture, fixtures, and equipment, including all of the furniture, fixtures (decor elements such as draperies, wall treatments, light fixtures), and equipment within foodservice areas.

Food Cost

The ratio of dollars of food purchased to revenue, expressed as a percentage. Unless certain paper products are directly related to the service of food (such as disposable plates, cartons, etc.), they should not be included in the food cost.

Food Court

A facility similar to a cafeteria foodservice operation. Looks similar to a multiunit foodservice operation in a mall, with separate, branded food and beverage stations, including national brands or brands that might be developed and are proprietary to the foodservice operator (or owner).

Foodservice Contractor

A foodservice company that specializes in operating restaurants (and catering and related food-and-beverage services) in cultural institutions, public attractions, corporate offices, schools or universities, convention centers, stadiums, hospitals, or similar facilities where the owner of the facility outsources this service. This is also known as a "contract foodservice operator."

Foodservice Operator

An independent foodservice provider/manager that operates an institution's restaurant, café, carts or kiosks, catering, and, possibly, vending machines. This entity collects all receipts (cash and charge sales from a restaurant, catering, and other program elements) and pays all direct costs and expenses (food, beverages, staff, staff benefits, operating supplies, insurance, etc.). A foodservice operator may be a national or regional foodservice contractor or local restaurant, caterer, or hotel.

Franchise Agreements

Contracts in which a franchisor grants a franchisee the right to use the franchisor's name and method of doing business. This is commonly done with branded concepts. Franchise and licensing agreements are very similar.

Franchisee

An operator who signs an agreement with a franchisor to use its branded concept.

Franchisor

Owner of a brand (branded concept).

Free Pouring

Dispensing of liquor by a bartender who manually (without any automatic alcoholic beverage dispensing equipment) pours and estimates the amount of liquor needed for a drink.

Front of the House

Comprises all areas accessible by guests, including the dining room, serving area (in the case of a food court or cafeteria), corridors, elevators, restaurants and bars, meeting rooms, and restrooms.

G&A

General and administrative expenses. See *administrative and general expenses.*

General Insurance

All types of insurance not related to employee benefits or extended coverage of a premises or its contents, including but not necessarily limited to security, fraud/forgery, fidelity bonds, public liability, food poisoning, liquor liability, use and occupancy, lost or damaged articles, and partners' or officers' life insurance.

Gratuity

An amount paid to foodservice employees by customers, either directly in cash, added to a charge sale ticket, or charged by a catering company to its customers. If charged by a catering company, the gratuity is then dispersed to the caterer's employees. A gratuity may be allocated to servers, chefs, bartenders, and others. A gratuity should not be confused with a service charge, which may or may not be paid to foodservice workers at their employer's discretion.

Gross Operating Revenue

Total payments received from customers (of restaurants, carts and kiosks, and catering) for goods and services.

Gross Profit

Gross operating revenue minus cost of goods sold.

Gross Sales

All revenue less (net of) applicable sales tax.

Hospitality Industry

Includes a wide range of businesses, each of which is dedicated to the service of people away from home; businesses that emphasize personnel's responsibility to be hospitable under the direction of hosts and managers of offered services.

Incentive Fee

An arrangement requiring a foodservice operator to take risk, usually in the form of placing a percentage of their fee(s) at risk based on the fiscal performance of the foodservice operations. Applies to management fee contracts.

Indirect Subsidy

Costs generally associated with the occupancy of a facility (such as rent, building and equipment maintenance, security, utilities, insurance, and property taxes) and the institution's management and administrative overhead charges (such as those for human resources, purchasing, and contract administration support).

Internal Catering

Catering services ordered, paid for, and presented by a cultural institution exclusively for its own meetings, member receptions, exhibition openings, development, dinners, and other functions.

Kiosk

A usually stationary and permanent (but sometimes temporary) small structure with limited food preparation and assembly equipment that has one or more openings (pass-out areas). Its purpose is to sell coffee, espresso, and related beverages (with the possible addition of cold bottled beverages) and baked goods and other snacks. Cooking is possible in a kiosk, and generally, local health departments place similar requirements on a kiosk as they do on a permanent restaurant or café.

K-Minus

A foodservice industry term for a foodservice facility that has no kitchen. In such a facility, food is usually brought from an off-premise central kitchen.

Labor-Intensive

Usually describes a large foodservice facility that, due to its layout and/or size in relationship to the number of customers it serves, requires more foodservice staff to operate, clean, and service, over and above what would be considered standard or normal.

Lease

This term usually applies to commercial restaurants that rent space from a landlord for a long fixed term with no cancellation by the landlord/owner except for serious breach or default, and with a capital investment by the foodservice operator. Leases are distinctly different from foodservice agreements and contracts commonly used in most cultural institutions.

Licensee

A foodservice operator that signs an agreement with a licensor or franchisor to use its branded concept(s).

Licensing Agreements

Contracts in which a licensor or franchisor grants a licensee or franchisee the right to use the franchisor's name and method of doing business. This is commonly done with branded concepts. Franchise and licensing agreements are very similar.

Licenses and Permits

Federal, state, and municipal licenses, including any special permits or inspection fees.

Licensor

Owner of a brand (branded concept).

Loose Equipment

Pots, pans, kitchen utensils, tableware (plates, glasses, flatware), and any other small, portable, or miscellaneous equipment not defined as FF&E. Also called *smallwares*.

Maitre d'

The dining room manager, who oversees the entire front-of-the-house operation. Usually employed at restaurants and catering events.

Management Fee

The fee, usually based on 3 to 5 percent of net sales, that a contractor receives for managing an account. This fee is usually received in addition to the G&A fee and would be considered a foodservice operator's profit.

Management-Fee Contract

Applies when an outside foodservice supplier operates an institution's foodservice program (which may include restaurants, catering, carts and kiosks, and vending machines) for a guaranteed fee (profit/income). Under this type of contract, profits, if any, from all food and beverage service operations are usually shared by the operator and the institution. The institution is normally responsible for any operating loss (also called *subsidy*). Operators will usually guarantee a mutually agreed-upon annual operating budget for food and beverage services based on a percentage of their guaranteed fee.

Marketing

Includes all selling and promotion efforts, direct mail, donations, souvenirs, advertising, public relations and publicity, and market research (surveys) covering restaurant and/or catering and special-events activities at a cultural institution. If foodservice is outsourced, this expense is usually shared between the foodservice operator and the institution.

Meeting Planner

Also called *event planner,* this person coordinates every detail of meetings and conventions. Meeting planners can be independent contractors or part of large corporations that have in-house meeting planning departments.

Menu Cycle

Repeating an established menu in a predetermined rotation. Generally, this rotation occurs every four or five weeks, but it can be done seasonally, as well.

Menu Mix

All menu items served in a restaurant, along with a count of how many servings of each item were sold during a given day-part. A menu mix tabulation lets an operator know what is selling and what is not; useful for menu planning purposes.

Minimum Guarantee

A minimum annual dollar amount, usually guaranteed against a percentage of gross operating revenue. Commission paid to client.

Net Profit

Gross operating revenue less all controllable and direct expenses, including (if applicable) any fees paid to a foodservice operator, before income taxes.

Net Sales

Gross sales less customer refunds, gratuities paid to foodservice staff, and the amount of discounts on restaurant/café/cart sales (i.e., if there is a 10 percent discount, 90 percent is included in gross sales).

Off-Premises Catering

Preparing food (and beverages) at a central kitchen and transporting products to the site where they will be served. Some caterers will transport food fully cooked and ready to serve, while others will pre-prepare meals before finishing them at a catering site.

On-Site Foodservice

Foodservice operations principally located in major civic institutions, such as cultural institutions, hospitals, schools and universities, corporate offices, penal institutions, nursing homes, military bases, and industrial sites; formerly known as *institutional foodservice.*

Outsourcing

The process in which the owner/administrator of a cultural institution, public attraction, corporate office, school or university, convention center, stadium, hospital, or similar facility contracts with an outside operator to manage its foodservices and/or its special-events department.

Payroll Taxes and Benefits

Payroll taxes include items such as Social Security taxes (FICA), state unemployment and state disability, social insurance (i.e., workers' compensation insurance, welfare, or pension plan payments), accident and health insurance premiums, and the cost of employee instruction and education, parties, and meals.

Per Capita Revenue

Total foodservice revenue (from a café, restaurant, cart or kiosk, or any other location where visitors purchase food and/or beverages) divided by the total annual attendance at a cultural institution. Total attendance is based on the institution's published public attendance numbers.

Point of Sale (POS)

Anyplace where food and beverage transactions take place; also, a computerized cash register system that allows restaurants and bars to preset menu prices and items.

Pre-Opening Expenses

Expenses incurred before a foodservice operation opens, including but not limited to travel, office, printing, advertising and promotion, training, and payroll for the pre-opening team.

Price-Value Relationship

Customers' perceptions of the value associated with a product relative to its cost.

Productivity

The measurable amount of work produced by foodservice staff.

Profit

Defined by a foodservice operator as total contribution from a program. A foodservice operator's profit expectation as a percentage of gross operating revenue is typically a minimum of 5 to 10 percent. Lower profit percentages are common in foodservice operations with higher gross operating revenue.

Profit-and-Loss Contract

An arrangement whereby a contractor is responsible for the financial operation of a foodservice, with risk placed upon the contractor to make money in the operation without a dollar subsidy from the client organization. Where a foodservice operator manages the institution's foodservice program under a profit-and-loss type of contract, the operator usually pays the institution a percentage of its total revenue. Also in practice is a hybrid of this arrangement whereby the client pays the contractor a flat weekly or monthly dollar amount.

Profit-Sharing Contract

An agreement that allows a foodservice operator and an institution to share in net profits.

Purchasing Specifications

Standards that ensure consistent quality as established by restaurant (and catering) management for the food, beverages, equipment, and supplies purchased; also called "specs."

Replacements

Replacements for linen, china, glassware, flatware, and kitchen utensils.

Request for Proposal (RFP)

When an institution is outsourcing or considering outsourcing and of its self-operated in-house departments, a request-for-proposal document is prepared that provides prospective operators with background and current information about the business opportunity.

Request for Qualifications (RFQ)

A request for qualifications asks companies to submit their qualifications to perform a described level of foodservice. RFQs are required only if administrators have no other way to prequalify prospective foodservice operators.

Restaurant

A type of foodservice operation that can range from cafés (with a small, limited menu) to large, full-service facilities offering sit-down table service and/or buffet service, cafeteria service, and/or self-service.

Revenue—Net

Gross food and beverage revenue (sales), net of sales tax.

Revenue—Restaurant

Includes food and (alcoholic) beverage sales/revenue, which are usually accounted for separately. Includes revenues from carts and kiosks.

Revenue—Special Events

Includes food and (alcoholic) beverage revenue, as well as revenue from other products and services that are sold users that host special events. This includes but is not limited to flowers, decor, entertainment, valet parking, and service staff.

Salaries and Wages

Includes salaries and wages, extra wages, overtime, vacation pay, and any commission or bonuses paid to employees.

Sales Mix

A listing by item (such as pizza, cappuccino, chocolate cake) and/or by sales group (such as entrees, beverages, desserts) of the number and/or dollar amount of menu selections sold during a specified period of time.

Self-Operation

This occurs when an owner of a restaurant, café, carts or kiosks, and/or food and beverage catering services operates with its own employees and management. The owner also handles all cash and accounts receivable and pays all bills and costs associated with providing these services.

Servery

The area in a cafeteria or food court where food is actually served to and picked up by customers.

Service Charge

A standard overhead charge by caterers, typically ranging from 15 to 18 percent of the food-and-beverage sales or total billing. Not a gratuity or tip paid to catering staff, unless a caterer so indicates this preference to a customer and designates that all or a portion of the service charge be so paid.

Smallwares

Hand held items necessary to support the preparation and service of food. They include such items as pots, pans, serving utensils, china, glassware, serving platters and bowls, and carving boards. *Smallwares* and *loose equipment* are synonymous.

Snack Bar

A walk-up, quick-service, permanent location that offers mostly prepackaged foods and beverages.

Special Events

Food- and/or beverage-related meetings, receptions, and affairs at a cultural institution. May be internal or external.

Special-Events Department

A department within a cultural institution that handles the master calendar for special events, facility rentals, approved caterers, and all details relating to the implementation and execution of special events. This department often also has the responsibility for advertising and marketing the institution as a venue for external special events. It is usually staffed and operated by the institution but is occasionally outsourced. Some cultural institutions outsource all or part of the management of this department to foodservice operators that are also providing in-house restaurant and catering services when catering services are provided on an exclusive basis.

Subsidy (Direct)

Dollar support from client provided to pay foodservices' direct and indirect operating expenses over and above the dollars received from sales (gross operating revenue).

Subsidy (Indirect)

Expenses that are absorbed by an institution and not charged directly to foodservice operations, often including the cost of utilities (gas, electricity, fuel, water, and sewage), trash disposal (Dumpsters), security, janitorial, telephone, repairs and maintenance to a building and FF&E, bookkeeping and accounting, and other similar expenses. These indirect costs and expenses are, in fact, costs and expenses associated with the operation of foodservices and should be tracked and/or so considered, especially by institutions that are planning new or expanded foodservices.

Tastings

Opportunities to view and taste menu items for restaurant, café, and/or catering service.

UBIT

Unrelated Business Income Taxation (see Chapter 10)

Utilities

Gas, electricity, fuel, water, waste removal, and sometimes telephone.

Vendor

Provider of food, beverages, equipment and/or non-consumable supplies. Also called *supplier.*

INDEX